Anna Sandak · Jakub Sandak ·
Marcin Brzezicki · Andreja Kutnar

Bio-based Building Skin

Anna Sandak
National Research Council of Italy
(CNR-IVALSA)
Trees and Timber Institute
San Michele all'Adige, Trento, Italy

Faculty of Mathematics
University of Primorska
Koper, Slovenia

InnoRenew CoE
Izola, Slovenia

Marcin Brzezicki
Wrocław University of Science
and Technology
Wrocław, Poland

Jakub Sandak
InnoRenew CoE
Izola, Slovenia

National Research Council of Italy
(CNR-IVALSA)
Trees and Timber Institute
San Michele all'Adige, Trento, Italy

University of Primorska
Koper, Slovenia

Andreja Kutnar
University of Primorska
Koper, Slovenia

InnoRenew CoE
Izola, Slovenia

https://doi.org/10.1007/978-981-13-3747-5

To our parents and children, who show us how beautiful the World around us is, and how important is to respect the Nature

Preface

Architecture is a very dangerous job.
If a writer makes a bad book, eh, people don't read it.
But if you make bad architecture,
you impose ugliness on a place for a hundred years

Renzo Piano

The trend to create sustainable buildings in addition to increasing environmental awareness has led to renewed interest in application of bio-architecture as an alternative to other construction techniques. Indeed, the expansion of bio-based product availability and their extensive utilization in modern buildings is one of the top priorities of the European Union's development strategies and its societal challenges. Accordingly, bio-based materials are considered a promising resource for buildings in the twenty-first century due to their sustainability and versatility. Unfortunately, architects and civil engineers are rarely trained in the proper use of wood and other biomaterials, including its use for building façades.

Façades are not considered to be a separate building component but an integral building element contributing to its performance and aesthetics. As an interface of the exterior and the interior, façades influence both: the area around the building and its internal space. New materials and smart technologies allow designers and engineers to propose innovative solutions to make buildings outstanding. Façades are the outer skin of a structure and are responsible for improving energy efficiency, while being both innovative and attractive. Consequently, traditional façade concepts have evolved. They have shifted from non-renewable to renewable, from barrier to interface, from invariable and static to responsive and dynamic, from passive single function to adaptive and multifunctional, from continuous to modular, from ordinary to customized, and from minimizing harm to being regenerative and restorative. These innovations are relatively simple to implement while designing and constructing new buildings, but their application in existing buildings is much more complicated. Around 35% of the EU's buildings are over 50 years old, and the restoration of residential buildings accounts for 65% of the renovation market.

It follows that a current challenge is upgrading the existing building stock in an effective and sustainable way. Retrofitting has become an important component of Europe's construction sector. New construction techniques, alternative materials, and innovative technologies enable us to retrofit old, and design new, buildings and structures to withstand the current and predicted impacts of climate change. Recovering buildings through renovation rather than demolition and reconstruction is a common practice but is a vital concern from a sustainability perspective.

This book is a result of the BIO4ever project funded by Ministero dell'Istruzione dell'Università e della Ricerca (MIUR) (RBSI14Y7Y4). The project was dedicated to filling the knowledge gaps related to the fundamental properties of novel bio-based building materials. During the BIO4ever project, the performance of 120 diverse bio-based building materials usable for building façades, that were provided by 30 institutes and companies was tested and evaluated. We hosted 24 researchers and visited several research institutes and universities establishing collaborations we hope to build in future.

This book is also an outcome of fruitful collaboration among researchers that had a chance to work together thanks to COST networking tools. Three COST Actions: FP1303 "Performance of bio-based building materials"; FP1407 "Understanding wood modification through an integrated scientific and environmental impact approach"; and TU1403 "Adaptive façade network" allowed us to work together, exchange ideas, and merge our expertise.

The "Bio-Based Building Skin" book was written to address a wide audience including architects, engineers, designers, and contractors. It provides a compendium of material properties, demonstrates several successful examples of bio-based materials applied as building skins, and provides inspiration for designing novel solutions. The state-of-the-art review and presentation of the newest trends regarding material selection, assembling systems, and innovative functions of façades are provided with an appropriate level of detail. Selected case studies of buildings from diverse locations are presented to demonstrate the successful implementation of various biomaterial solutions, determining unique architectural styles and building functions. Analysis of the structure morphologies and aesthetic impressions related to bio-based building façades is discussed from the perspective of art and innovation. Essential factors influencing material performance are argued from various perspectives, including aesthetics, functionality, and safety. Special focus is directed on assessment of the performance of façades throughout the service life of a building, including end of life. This book provides technical and scientific knowledge and contributes to public awareness, by providing evidence of the benefits to be gained from the knowledgeable use of bio-based materials in façades.

In reference to Renzo Piano's statement quoted above, we hope that our book will highlight good architecture, will inspire your future research and projects, and will be worth reading…

San Michele all'Adige, Italy/Izola, Slovenia — Anna Sandak
Koper, Slovenia/Izola, Slovenia — Jakub Sandak
Wrocław, Poland — Marcin Brzezicki
Koper, Slovenia/Izola, Slovenia — Andreja Kutnar
September 2018

Acknowledgements

No duty is more urgent than that of returning thanks
James Allen

We would like to acknowledge all the comments and feedback we have received, during the preparation and writing of this book from individuals, firms, and organizations. This book is a result of the BIO4ever project that will be impossible to conduct without the funding from Ministero dell'Istruzione dell'Università e della Ricerca (MIUR) (RBSI14Y7Y4). We would like to thank CNR-IVALSA for opportunity to perform this research during last years and small BIO4ever team: Marta Petrillo and Paolo Grossi from the Laboratory of Surface Characterization for their exceptional work. The authors gratefully acknowledge the European Commission for funding the InnoRenew CoE project (Grant Agreement #739574) under the Horizon2020 Widespread-Teaming programme and the Republic of Slovenia (investment funding of the Republic of Slovenia and the European Union of the European Regional Development Fund) allowing us to continue our research in their inspiring and vibrant team. Great thanks to all colleagues from both institutions for their support and help, and in particular to Dean Lipovac, Mike Burnard, and Liz Dickinson for their valuable comments and English edits.

This book is also an outcome of fruitful collaboration among researchers that had a chance to work together thanks to COST networking tools. Three COST Actions: FP1303 "Performance of bio-based building materials"; FP1407 "Understanding wood modification through an integrated scientific and environmental impact approach"; and TU1403 "Adaptive façade network" allowed us to perform investigation within outstanding research teams. We would like to thank our Norwegian colleagues: Ingun Burud, Thomas Thiis, Lone Ross Gobakken, and Peter Stefansson for brainstorming during modelling of weather dose. Our thanks to Wim Willems for inspiring discussion about wood hygroscopic properties and "façade hunting" afternoon when we discovered many bio-based buildings in the Netherlands. Charalampos Lykidis for trials with thermally modified wood. Athanasion Dimitrou for measuring of thousands of samples from round robin test.

Dominika Janiszewska for first experiments on liquefaction of modified wood. Magdalena Kutnik and all FCBA teams for their training and assistance while working with termites and fungi. Our greatest thanks to Veronika Kotradyová and Agnieszka Landowska for inspiring discussion and our first experiments regarding material–human interaction.

Evaluation of material performance will be impossible without support of industry and academia. Hundred and twenty materials provided by 30 partners from 17 countries were intensively tested, characterized, and modelled. Our greatest thanks to: ABODO (New Zealand), Accsys Technologies (the Netherlands), Bern University of Applied Sciences (Switzerland), BioComposites Centre (UK), Cambond (UK), Centre for Sustainable Products (UK), Drywood Coatings (the Netherlands), Eduard Van Leer (the Netherlands), FirmoLin (the Netherlands), Houthandel van Dam (the Netherlands), ICA Group (Italy), Imolalegno (Italy), Kebony (Norway), KEVL SWM WOOD (the Netherlands), Kul Bamboo (Germany), Latvian State Institute of Wood Chemistry (Latvia), Lulea University of Technology (Sweden), Novelteak (Costa Rica), Politecnico di Torino (Italy), Renner Italia (Italy), Solas (Italy), SWM-Wood (Finland), Technological Institute FCBA (France), Tikkurila (Poland), University of Applied Science in Ferizaj (Kosovo), University of Gottingen (Germany), University of Life Science in Poznan (Poland), University of Ljubljana (Slovenia), University of West Hungary (Hungary), WDE-Maspel (Italy). We do hope that this project was just beginning of our collaboration and we are looking forward to continuing it.

The photographic material used for this book was collected during different field study trips financed from grants from Ministry of Science and Higher Education in Poland, granted to Wroclaw University of Science and Technology, Faculty of Architecture and by COST Action TU1403 "Adaptive façade network". Our greatest thanks to all companies, institutes, and friends who additionally supported us with images illustrating beauty but also problems while implementing biomaterials. In our book, we tried to be objective and to show good inspiring architecture, since "*There is no truth. There is only perception*". Gustave Flaubert

September 2018

Contents

Chapter 1
State of the Art in Building Façades

Everything is designed. Few things are designed well.
Brian Reed

Abstract This chapter presents a portfolio of building materials suitable for façades. It describes the relationship between material type, building element, façade, and the entire building structure. Traditional façades based on static components, as well as adaptive concepts able to interact with changing environmental conditions, are briefly described and illustrated with pictures. Climatic design principles, biomimicry, and bioinspiration in architecture are introduced with the purpose of inspiring future developments.

The function of façades in architecture and the big portfolio of protective layers developed by nature (skin, membranes, shells, cuticles) share several similarities. In nature, skin is the largest organ that protects the body from external invaders. Skin is a multitasker performing several functions critical for health and well-being of organisms. Built from several layers, skin protects, regulates, controls, absorbs, maintains, senses, and camouflages. The analogies between functions of the building façades and animal skin are presented in Fig. 1.1. Building façades partly define architectural characteristics of structures and act as a shelter and space for human activity (Gruber and Gosztonyi 2010). They provide UV, moisture, and thermal defence, as well as protection from dirt, micro-organisms, and radiation. Façades communicate by transferring information—they are capable of exchanging and storing energy, heat, and water.

Since the first buildings were constructed, façades have been separating two environments: external and internal. To maintain constant internal climatic conditions, façades had to counteract the influence of various external environments depending on the given climate zone. In hot and humid zones, they provided protection against the sun radiation and allowed for the flow of cooling night breezes. In temperate climates, façades had to adapt to seasonal changes. In harsh north environments, façades were mainly designed to protect against the winter cold. This affected not only the construction material used but also the shape and configuration of windows, building orientation, and the heating strategy.

A. Sandak et al., *Bio-based Building Skin*, Environmental Footprints and Eco-design of Products and Processes, https://doi.org/10.1007/978-981-13-3747-5_1

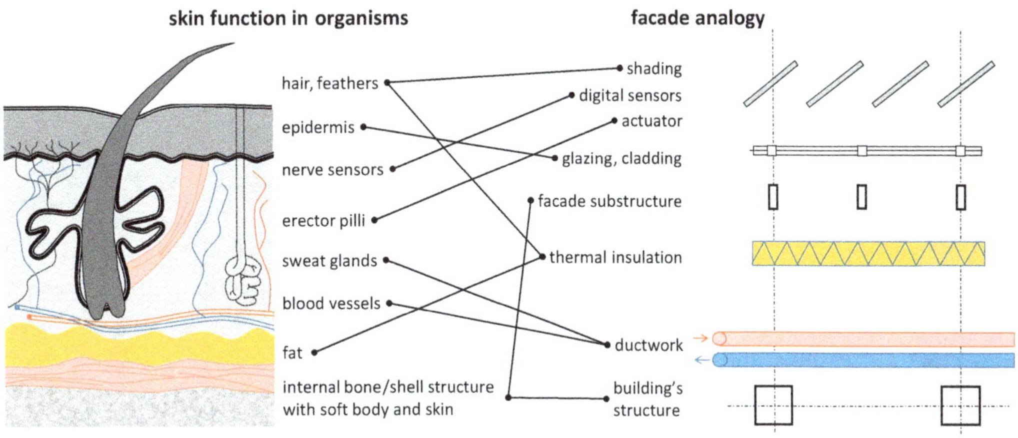

Fig. 1.1 Analogy between animal skin and building façade

In addition to possessing the obvious structural and protective functions, façades also needed to be durable. The most robust materials (e.g., stone) were usually the most expensive and most difficult to acquire. The scarcity of stone led to the development of brick, where the areas abundant in clay were available. Simple adobe brick stemming from dry climates was gradually replaced by the fired brick coming from the north, as this type of treatment provided a long-lasting waterproof layer. However, before the invention of masonry, from the very beginning of architecture, buildings were constructed of wood and other bio-based materials.

1.1 Structure–Façade–Element–Material

A shelter is usually defined as an enclosed space—a space that is somehow delimited from the surrounding environment to enable control over the internal microclimatic conditions. Such a space is the basic spatial element of habitation in many cultures and climate zones. With time, the number of rooms gradually increased, and each space supported a separate activity, such as socializing, sleeping, cooking, storage, animal raising, and cattle breeding. The type of enclosure delimiting the room depends on the exterior climatic conditions, available resources, and lifestyle. In nomadic tribes, light and portable shelters were developed as low-weight fast-to-erect solutions (Fig. 1.2). Those included two clearly separated elements—the load-bearing skeleton and the protecting envelope. With the onset of the static settlements, weight of the shelters was no longer a concern, while their robustness and ability to protect from enemies became a priority. This important change facilitated the invention of the "wall" that could be defined as a multifunctional element providing both the enclosure and load-bearing properties. The following chapter provides a brief overview of the gradually decreasing scale of complexity related to buildings.

Fig. 1.2 Yurts in Kyrgyzstan. Image courtesy of Aleš Oven—InnoRenew CoE

1.1.1 Structure

In human dwellings, two basic principles were developed to construct the envelopes: flexible and less permanent façades represented by tents for mobile use used in arid climates by nomadic tribes, and fixed solid walls designed to last. Walls were generally constructed from locally available materials since the immediate surroundings often provided technical solutions (Knaack et al. 2007). Initially, mud bricks and stone were used. With the spread of humankind further north (continental Europe), new requirements arose for the existing enclosures. Namely, they had to provide an improved insulation from heat and cold as well as better heat conservation properties (air tightness protecting from the heat loss through ventilation). With the scarcity of stone in available form (i.e., fewer available pebbles—more material had to be derived from a quarry), wood became a sensible alternative, as it offered insulation properties surpassing that of stone and proved to be much easier to acquire. This resulted in the widespread development of timber-based wall systems used almost in all climate zones where wood was available.

Timber became a main material in early Medieval cities and was later gradually replaced by stone used in fortifications following the rapid advancement of artillery. Still, timber buildings were not only built within the boundaries of fortified city walls but also on the outside. The latter case illustrates a deliberate strategy, as those buildings could be set on fire just before the enemy invasion.

1.1.2 Façade

The term "façade" generally refers to the external surface of the wall. Sometimes, however, the term is reserved to name only the frontal part of a building (e.g., a theatre overlooking the plaza has just one façade). The term comes from post-classical Latin facia (meaning "human face"). Initially, a façade was simply a by-product of a material type used in wall construction, as external and internal surfaces were no different. Gradually, with the increased significance of aesthetics, the external layer of human shelters acquired refinements that were pleasing to the eye. This lead, together with the development of geometry and mathematics, to the expansion of architecture, which became characterized by different building styles, proportions (e.g., the golden ratio), and classical orders. Façades were also able to advertise the power and prestige of building occupants long before other means of communication, such as writing, were invented. Throughout history, façades functions have been changed and upgraded in response to emerging technologies and materials. They continuously evolve and adapt in order to satisfy the changing demands of occupants (Capeluto and Ochoa 2017).

Humankind developed numerous materials to be used as façade coverings. Some of them originated directly from the wall material (e.g., timber, stone, brick), while others were developed deliberately as covering materials and were intentionally different than materials used in wall construction. Such materials are, for example, plaster, daub, ceramic tiles, and recently developed curtain walls made of steel, aluminium, and glass. The development of these materials was primarily motivated to seal the wall from the air penetration (this holds especially true in the case of plaster that seals the wall but remains vapour permeable) but also to protect a relatively fragile structural material from harsh environmental conditions. This external finish is also intended to either totally block (e.g., curtain wall) or at least control (e.g., in so-called ventilated cladding) potentially hazardous water ingress. Often, the external façade cladding material is less durable than the structural material; thus, occasional refurbishment of façades is foreseen to maintain the overall building durability.

1.1.3 Element

As log houses were probably the first biomaterial-based permanent structures erected by humans, logs represent the first bio-based building elements. Gradually, with the development of tools and building techniques, the framed structures were developed. Those used timber members in different structural components: as beams, posts, columns, bracing elements, rafters, etc. In this type of structure, timber cladding elements (e.g., profiled boards, laths, and wood shingles–shakes) are also considered to be bio-based façade elements. The advent of industrialized production allowed for more efficient timber manufacturing techniques, while the advancement in chemistry provided various man-made binders. This facilitated the production of bio-based materials composed of both full-featured (laminated timber, cross-laminated timber) and previously rejected waste (by-products) materials (chips, sawdust or even stems and leaves). Consequently, a wide variety of elements arose, including block boards, particle boards, oriented strand board (OSB), and—most recently—natural fibre-reinforced polymer composites.

1.1.4 Material

As experience shows, properly handled structural bio-based materials proved to be durable and robust much longer than expected (some timber-framed structures show the astonishing lifespan of a few hundred years). However, bio-based materials used as external cladding face the most severe climatic conditions as they are fully exposed to the outdoor environment. In a temperate climate, the façade is exposed to daily frost and thaw cycles in spring and autumn and to large seasonal fluctuations in UV levels. Considering this, it seems that the protection from the environmental influences is the most important issue in the application of bio-based materials. Controlled ageing of the bio-based materials has thus become a crucial engineering challenge.

Natural Stone

Stone is a natural substance, a solid aggregate of one or more minerals. It is the main building material of the Earth's outer solid layer, the lithosphere. Due to its availability, the stone has been used by both human and prehuman species as a tooling and building material since the advent of civilizations (approximately 12,000 years ago). Stone was initially obtained by collecting and later by mining. In the construction industry, stone can be used in its natural form as a pebble (boulder) or it can be mechanically transformed into a desired form, for instance, a wall element (ashlar). Those basic elements can be either assembled loosely (dry stone wall) or glued using the mortars or plasters.

Stone has a high heat capacity and is generally exceptionally durable, although this varies somewhat depending on the type of rock: for example, while granite is very durable, marble is prone to damage induced by chemical agents. The stone cut in thin slices (approximately 2 cm) is commonly used in buildings and is recognized as a highly durable cladding material. Initial high cost of production is usually justified by low maintenance costs. However, the dramatic increase in deterioration in the built heritage has been observed during the past century due to climate change and increased environmental pollution (Siegesmund and Snethlage 2014).

Concrete

Concrete is a composite material consisting of a fine and coarse aggregate (i.e., sand and gravel mixed in various proportions) bonded together with a fluid cement (i.e., cement mixed with water forming a so-called cement paste). Due to a series of chemical reactions, concrete hardens over time. The Portland cement type is usually used to manufacture concrete, but other binders might also be used, for example, lime-based binders or asphalt. In the production stage, concrete has a form of a slurry that is poured into the formworks typically made of timber or prefabricated elements. As a consequence of certain chemical processes, concrete forms a stone-like material that can differ in strength and other properties depending on the content of the mixture that was used to manufacture the slurry. Concrete is commonly labelled as "artificial stone".

Because of the similar Young modulus in concrete and steel, the steel bars—also called "rebars"—are widely used to reinforce concrete. Concrete is therefore used in the compression zones, while the steel reinforcement is used in the tensile zones to provide tensile strength. Concrete and stone (the aggregate used to manufacture concrete) are similar in weight; however, reinforced concrete is heavier due to the added steel inserts. Concrete has a moderate environmental impact. Bribián et al. (2011) compared the environmental impacts of various building materials, while taking into account the manufacturing, transport, construction, and demolition of buildings. The functional unit applied was 1 kg of the material. The primary energy demand of concrete was calculated to be approximately 1.1–1.7 MJ eq/kg. The higher values are reported for reinforced concrete (Bribián et al. 2011). Cement, on the other hand, has a higher primary energy demand, 4.2 MJ eq/kg, due to the high energy use and generated pollution in the production phase. Although the proportion of cement in the concrete is relatively low, approximately 1/7th of the total mass of the concrete, it considerably contributes to the overall environmental impact of concrete. It should be emphasized that the results could be different if calculated per 1 m^3 of the material, especially when accounting the life cycle assessement (LCA) for materials with different physical properties (Bribián et al. 2011).

Ceramics

A ceramic is a non-metallic solid material comprising inorganic compounds (metal, non-metal, or metalloid atoms). Ceramics can have a crystalline or non-crystalline internal structure (e.g., glass). Building ceramics (e.g., fired bricks) represent a certain range of crystallinity, where the atoms or molecules are arranged in regular periodic microstructures. Ceramics are produced from different types of raw ceramic materials (e.g., clay, ash, different chemical forms of silica) using high temperatures ranging from 1000 to 1600 °C in the process called firing. Ceramic materials are hard, and strong in compression but also brittle, and weak in shearing and tension. As the material is not amenable, it generally has to be shaped during the production stage because of its inherent brittleness.

Ceramics have high heat capacity, yet, due to its porous structure, lower than concrete and stone. Glazed ceramics are very durable and resistant to chemicals. They have been used in buildings since the beginning of the civilization, that is, as soon as firing kilns were invented. In the past, brick was extensively used for the load-bearing construction but is nowadays commonly replaced by concrete and steel (brick is prone to failure in harsh conditions, such as earthquakes). In the present day, ceramic products are commonly used as the building external cladding material and are considered to be safe and durable. However, their environmental impact is relatively high, especially when covered with glaze. For example, the primary energy demand of ceramic tile is approximately 15.6 MJ eq/kg (Bribián et al. 2011).

Glass

Glass is a non-crystalline amorphous solid. It belongs to the group of non-crystalline ceramics being formed from melts. The basic melt contains 75% of silica (SiO_2), lime, and other ingredients (Na_2O, Na_2CO_3) along with several minor additives. In the temperature of approximately 1600 °C, the vitrification process occurs, and silica molecules are positioned randomly without a crystalline structure. After a period of cooling, the transparent glass is formed. Glass, as all ceramics, is brittle, hard, strong in compression but weak in shearing and tension.

Glass should be formed in melt form by casting or blowing. Around the beginning of the previous century, a series of flat glass producing technologies were invented, later topped by the float process developed by Pilkington in 1958. Flat glass can be machined, cut, and ground to the desired form, and it can be hot and cold bent. In the modern building industry, glass is commonly used to glaze windows in the form of insulating glass units (IGU) or employed as a safety glass that holds together when shattered. The typically used interlayers holding the glass in place are polyvinyl butyral (PVB) or ethylene-vinyl acetate (EVA).

Glass is resistant to numerous damaging chemicals. The material is commonly used to manufacture containers for a selection of aggressive chemical substances (acids and alkali). It was initially used solely as a material for window glazing, but after the invention of the curtain wall, it is commonly applied as a whole façade building's cladding for both visual and non-visual (meaning the transparent and opaque) parts of the façade. Since glass requires abundant levels of energy to be produced, it generates a high environmental impact. When calculating the impact from the phase of manufacturing to the time of building demolition (cradle to grave), the primary energy demand of glass is approximately 15.5 MJ eq/kg. The environmental impact, however, would be significantly lower if the calculation would include recycling (cradle to cradle), as glass has a high recycling potential (Bribián et al. 2011).

Metals

Metals are a group of metallic chemical elements, typically lustrous substances, characterized by a high electrical and thermal conductivity. Metals in pure form are rare in Earth's lithosphere; thus, they have to be extracted from the naturally occurring minerals—usually the ores. Metals and their melts were crucial in the technological development of human civilization (Bronze Age, Iron Age) and are currently widely used.

Metals appeared in the building industry in the mid-nineteenth century, following the onset of the Industrial Revolution. They gradually replaced the framed timber structures because of their greater strength and fire resistance. Currently, steel is used both as a structural, load-bearing material and, in the form of the thin sheet, as the façade cladding.

Metals (e.g., steel) are characterized by high compressive and tensile strength but have to be protected to slow down the oxidation process that might eventually destroy metal elements. They are prone to the damage resulting from a wide range of chemical agents, including acids and alkali. Metals are heavy—those used in buildings are approximately 4 times heavier than concrete and stone. The primary energy demand of metals, when considering the LCA calculation production, transport, construction, and demolition, ranges from 24.3 MJ eq/kg for reinforced steel to 136.8 MJ eq/kg for aluminium (Bribián et al. 2011). Due to the relatively high purchase cost, metals are commonly collected at the end of their life cycle and recycled to create new products, which reduces their environmental impact.

Plastics

Plastic is a material consisting of a wide range of synthetic or semi-synthetic organic compounds. Plastics are malleable and can be moulded into solid objects. They are typically organic polymers of high molecular mass and usually contain a variety of substances. Plastics derive from petrochemicals; however, variants that are made from renewable materials are also available.

Plastics were invented at the beginning of the twentieth century and became widely used a few decades later in many fields, including the building industry (e.g., polyvinyl chloride (PVC), polycarbonate (PC), hollow polycarbonate). Plastics are usually characterized by a high water resistance, lightness, and low structural strength; however, those parameters can vary depending on the composition of the polymer matrix. Plastics are combustible, and for the safety reasons, their use in buildings should be restricted (the most resistant plastic polymers can withstand the temperature of maximum 150 °C). Plastics are also prone to the UV radiation, which is considered as the most degrading agent for the majority of plastics manufactured today.

In the building industry, plastics are commonly used in a variety of forms and shapes. They are typically used in the PVC window frames that replaced the timber frames due to their lower cost of maintenance and good heat insulation. Many polymers are used in building industry in the form of thin coatings and paintings. Environmental impacts of plastics are among the highest when compared to other building materials. For example, the primary energy demand of polyvinyl chloride is approximately 73.2 MJ eg/kg. This is due to the high energy demand during the production phase as well as due to the high water consumption (one of the largest among building materials, exceeding 0.5 m^3 for a kg of the finished product) (Bribián et al. 2011).

Coatings and Paintings

Coatings are manufactured from a wide range of materials, usually plastics and metals that are deposited on a base material in a very fine and thin layer. Although the coating layer is thin, it can substantially modify the properties of the base material by changing some of its characteristics. For example, a few nanometres thick layer of a metallic coating applied on glass can substantially modify glass transparency and, consequently, its heat conduction.

In general, paintings and coatings are applied for functional and decorative purposes. Coatings may be used to improve certain physical characteristics, such as fire resistance, sealing, or waterproofing but could also be used to stain the base material (e.g., wood stains). Metals are coated to prevent oxidation and slow down the corrosion.

Application of different materials on building façades is presented in Fig. 1.3.

1.2 Standard Façades Systems

Façades are generally considered as static elements of buildings. This applies to the solid walls that fulfil both the load-bearing and protective function (the external face of those walls becomes the façade). It also applies to the most recent skeleton post and beam structural systems that are clad by lightweight curtain glazed walls. Initially, in all buildings, the internal conditions were maintained via natural means, for example, by ventilation and heat/moisture exchange through the envelope. They were open systems. The control of the environmental conditions was executed manually, simply by adjusting window openings for air exchange or by pulling the curtains for sunshading (Fig. 1.4). The parameters of the façade remained constant (e.g., the wall's U value or other structural properties) from the moment the façade was erected.

Especially in harsh climatic conditions, the maintenance of internal comfort required a lot of effort; thus gradually, some modifications were done to improve this process. In arid climates, the overhangs and the mass of the building were used to reduce the diurnal changes in interior temperature. In temperate climates, cavity wall and double glazing appeared as one of the best means for reducing the loss of heat. Buildings were temporarily sealed to reduce heat ingress/escape. Night-time cooling was practised in hot climates as a mechanism of seasonal ventilation and temperature lowering. In temperate climates, insulation was added in the winter.

The issues of the façade functions were addressed previously in numerous sources, including the iconic Façade Construction Manual (Herzog et al. 2004) and Timber Construction Manual (Herzog et al. 2012). Both manuals include one of the most comprehensive engineering resources for architects and designers. In 2014, a book titled Modern Construction Envelopes (Watts 2014) was published by Birkhäuser, containing a chapter dedicated to "timber walls". One of the most recent books titled: Building Envelopes: An Integrated Approach published in 2013 by Princeton Architectural Press provides instruction for designing building envelopes that are both visually appealing and high-functioning (Lovel 2013).

Fig. 1.3 Building façades with different materials: assumption Cathedral, Koper; Büro- und Produktionsgebäude in München/Kurt Tillich, tillicharchitektur, München; Champalimaud Centre for the Unknown/Charles Correa Associates; Skattekvartalet-Tax Administration/Narud Stokke Wiig; European Solidarity Center/FORT—Wojciech Targowski; St. Jakob-Park Basel Stadium/ Herzog & De Meuron; Museum der Kulturen/Herzog & De Meuron; Elnos commercial centre, Brescia

Fig. 1.4 Façade with integrated classic windows for direct air exchange (DKV Insurance Company HQ, Cologne, arch. Jan Störmer Architekten, 2005)

1.2.1 Closed and Open Systems

The invention of air conditioning by Willis Haviland Carrier in 1902 and introduction of heating, ventilating, and air-conditioning (HVAC) systems offered several options in microclimate regulation of buildings. Full mechanical microclimate regulation includes temperature, humidity, air circulation, and air cleaning. In buildings, the air exchange is powered by mechanical fans and provided via air ducts. This technology revolutionized the world building industry, starting initially in the USA in 1920. It enabled the manufacture and maintenance of microclimates in high-rise buildings with limited accessibility to windows. This technology is also supposed to facilitate the great migration to the so-called Sunbelt in USA (areas south of the 36th parallel). The technology creates an artificial environment that is maintained by fans, coolers, heaters, and dehumidifiers (generally speaking: compressor systems) channelling air through ducts in full insulation from external conditions. In this system, the façade becomes a tight impermeable barrier separating two environments, since the opening of the windows would disturb the constant indoor air conditions. Initially, HVAC systems seemed to be the solution to maintain certain microclimates in buildings, but eventually numerous drawbacks began to emerge. Filters used in air-conditioned buildings were imperfect, and some mould spores were admitted inside the ducts. As mould does not need the daylight to grow, it flourishes inside the warm and humid ducts producing toxic and dangerous substances subsequently inhaled by the building occupants. Stagnant and kept at the constant temperature, water facilitates the growth of various dangerous germs, such as Legionella pneumophila. On one occasion, dangerous bacteria migrated from the water present in the humidifying system to the air circulating in the ducts and caused the death of 29 elderly men that contracted severe cases of pneumonia (this event is currently labelled "the Legionnaire's diseases outbreak") (McDade et al. 1997). The outbreak received significant media coverage, drawing the public's attention to the issues of the indoor air quality and so-called sick building syndrome. Although the mechanical systems proved to be very effective in maintaining the proper temperature and humidity, the initial system version raised a multitude of questions about the hygiene and health conditions in buildings.

1.2.2 Mixed-Mode Systems

In contrast to isolated environments, open systems are characterized by outside conditions that help maintain occupant comfort and positively influence the interior climate. This is achieved by opening the building façade to the external environment in certain periods of the year. Therefore, contemporary ventilated façades had to work in mixed modes. The first one, an air exchange mode, is achieved by opening the windows (possible even in high-rise building thanks to the use of double-skin façades), while the second is the air-conditioning mode, used in the presence of extreme external conditions (either too hot or too cold). In the latter, the building is sealed, and the air is artificially exchanged by the mechanical system via ducts. The mixed-mode systems (combining both modes) proved very effective and were successfully introduced in numerous buildings. The additional advantage of such systems, apart from the improvement of internal air quality and substantial reduction of energy bills, was the inclusion of the occupants in the process of decision-making by allowing them to open windows, adjust the blinds, and regulate internal temperature to a certain extent. By the first decade of the twenty-first century, previously used fully mechanical systems were overshadowed by new advancements. One example might be the Commerzbank in Frankfurt am Main

Fig. 1.5 Commerzbank in Frankfurt am main (arch. Foster and Partners, 1997)

(a) (b)

Fig. 1.6 Capricorn Haus Façade (arch. Gatermann & Schossig, 2005). **a** View of the building's façade for the outside, **b** view of the façade module from the inside; see decentralized air-condition system installed in the façade module

(arch. Foster and Partners 1997) that can be naturally ventilated for approximately 60% of the year (Fig. 1.5). The systems have progressed even further towards the decentralized air-condition system in the Capricorn Haus Façade (arch. Gatermann & Schossig 2005), where every façade module is responsible for the climate regulation of an adjacent room (Fig. 1.6) (Brzezicki 2007).

1.3 Climatic Design

Climatic design is a set of design methods and principles used with the purpose of capitalizing on the advantages of climatic conditions surrounding buildings. In the paradigm of "climatic design", buildings try to gain the most from their local environment just like organisms rely on the resources provided by nature. The general rule is to work with the climate and not against it, by using all the available methods to provide comfort in the building with the minimized use of primary energy either for heating or cooling. Since climates are diverse, it is not surprising that specific climates require unique design solutions that could be generally divided into two groups: heat accumulating and heat rejecting techniques. Here, the influences of air humidity have to be taken into the account because, as the psychrometric chart shows, the same air temperature (dry-bulb temperature) proves to be either acceptable or unacceptable depending on the relative air humidity measured by means of a wet-bulb temperature (Lechner 2008).

The effectiveness of certain passive methods of internal climate regulation depends on the climate conditions. Evaporative cooling—the heat rejection technique—is effective only in a relatively dry climate (basically below 40% of humidity, the dryer the better), while the increase of the thermal mass to accumulate heat works in both dry and humid climates (Fig. 1.7a). The latter technique, together with the so-called night cooling/purging, is effective even in high daytime temperatures (up to 42 °C, at approximately 60% of humidity). Natural ventilation effectively lowers the temperature in buildings if, of course, the external air temperature is lower than the desired internal temperature (Fig. 1.7b). When the external air temperature is too high, the ventilation provides the hygienic air exchange but also introduces a lot of heat energy that can increase the interior temperature. In many warm climates, a shading is very effective in preventing the heat build-up in buildings.

Depending on the exact geographical location, wind is also used to lower the temperature and provide the ventilation in buildings, especially in locations where wind direction is constant, and the pressure difference can be used to channel the air through the building. So-called wind towers are used to catch and direct the stream of fresh air, and air tunnels are used to lower the air temperature.

In temperate climates, a different strategy is required as the weather conditions change significantly with seasons. In the hot part of the year, all the warm climate methods can be used, while in the cold season the heat accumulation techniques are more appropriate, including passive solar gains due to the greenhouse effect (e.g., by using glazed atria or conservatories). In cold climates, heat insulation is required to slow down the heat escape during the long and cold winter season.

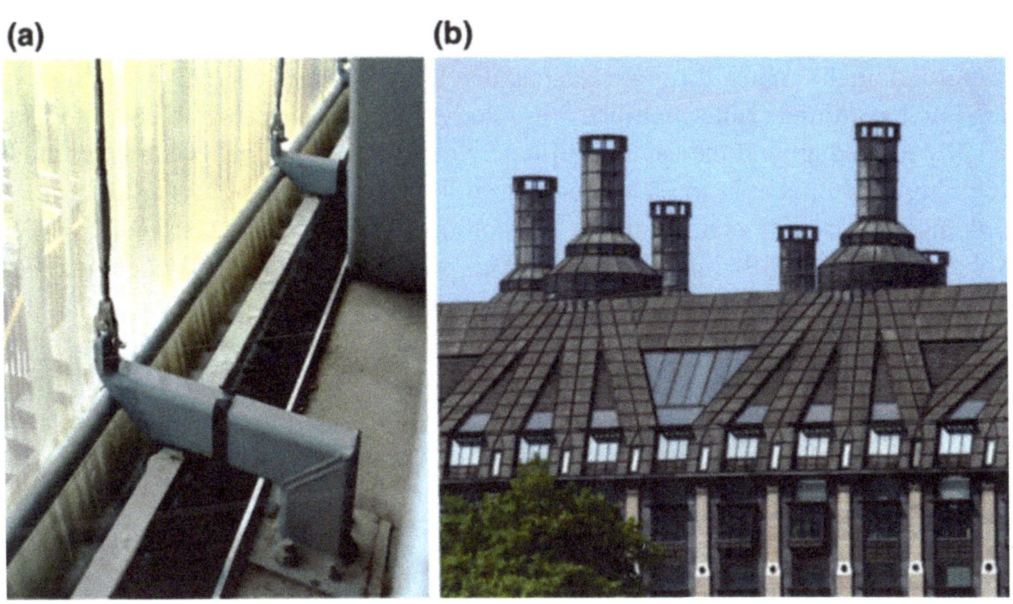

Fig. 1.7 Examples of passive methods of internal climate regulation. **a** Evaporative cooling and **b** natural ventilation

Fig. 1.8 Different buildings architecture all over the world

Different passive techniques were applied in diverse climates by using various building materials, depending on the availability. The use of those materials is usually a part of a local vernacular architecture or building traditions (Fig. 1.8). In vernacular architecture, many typological modifications are present, including the decreased window sizes, installation of an extra glass layer or timber shutters.

In hot and arid climates, building mass usually results from using stone or adobe (sun-dried brick). When those are scarce, dwellings can be "carved" in bedrock or dug into the ground. Tunisian tribes of troglodytes live in the artificial caves that are radially located around the central artificially dug pit. Traditional building methods (e.g., high accumulation mass and night ventilation) are currently also used in the building industry, for example in the Torrent Pharmaceuticals research centre (arch. Abhikram 1994–1999) in Ahmedabad. This building regained all constructional costs from the electrical savings in 13 years of operation, thanks to its passive downdraft evaporative cooling system (PDEC) and minimal air conditioning. In hot and humid climates, nature provides building materials in the form of timber (unmachined timber in a form of palm tree trunks) or plant leaves. These are used to erect lightweight and well-ventilated structures that achieve maximum cross-ventilation and convective air flow. Timber and plant leaves are also used to construct ventilated roofs with the deep eaves to provide shading and rain protection. Building openings (windows and doors) are in this case facing the predominant wind direction (Samuel et al. 2017). In temperate climates, different construction techniques have been developed. Stone is frequently used, but the fired brick becomes an alternative as a robust and water-resistant façade material. Depending on the availability of wood, either solid (in log houses) or framed timber structures are used. Both are easy to erect and relatively long-lasting. In the north, timber is becoming increasingly important due to the insulation potential resulting from its cellular microstructure. Framed timber structures were often filled with the lower-grade material: either with a wattle and daub or by the mix of clay and leaves providing insulation layer protecting against the heat escape in the winter. This vernacular technique evolved with time and is nowadays commonly used together with non-combustible modern insulation materials, such as the mineral or rock wool. In the mid-nineteenth century in northwest Europe, a cavity wall was developed as an effective heat insulation technique. It became widespread in the 1920s, and since the 1970s, it has been used in the wall cavities (e.g., polystyrene). Glass has become an important passive material in climatic design, since it is useful in conserving heat through the greenhouse effect. Glass production techniques, originating from the start of the twentieth century, are able to manufacture large thin glass sheets used to construct building envelopes. Even though metals (e.g., steel and aluminium) are not directly connected to the climate design, they are gaining importance by replacing timber-framed structures. Robustness and water-resistant qualities of metals are especially beneficial to glazed structures.

1.4 Biomimicking and Bioinspiration in Architecture

The systems found in nature are a valuable source of inspiration in many areas. The concepts of biomimicking and bioinspiration are being adopted by scientists and researchers from various fields, including structural engineering, robotics, medicine, and materials science. In the last years, the potential benefit derived from natural solutions has become appealing to the field of sustainable architecture. Biology serves as the initial basis for comparison and understanding of biomimetic principles. Then two approaches might be used to transfer the information: solution based (bottom-up approach) and problem based (top-down approach) (Gruber 2011). For the practice and realization of future urban designs, the architectural and engineering aspects of biomimetics have been far more distinctly developed (Pohl and Nachtigall 2015). In many cases, this approach complies with the general principle of so-called climatic design, which is one of the best approaches to reduce energy consumption in buildings. This analogy is hinting at how to design buildings that are using materials and energy optimally. In fact, the message for architecture that emerges from observing nature is "less materials, more design" (Pawlyn 2016).

1.4.1 Animals Inspiring Design

A paper published by Michael Davies titled "Wall for all seasons" presented a complex multi-layered façade envelope that is regulated electrochemically (Davies 1981). Its structure was analogous to the morphology of organic covering tissues (the skin), and the thermal regulation was comparable to biological mechanisms. Outside temperature and air humidity influence physical parameters of this "polyvalent" wall, that—besides sealing and insulation—provides other functions, such as light transmittance, thermal conductivity, and vapour permeability. The wall's layered structure fulfils different functions: external "weather skin" protects against rain and cold, while two "sensor and control logic" layers gather information about the status of the external environment. Layers equipped with micropores regulate the intensity of gas (vapour and air) infiltration. The presented system included a photovoltaic layer, which enables energy generation. The author also presented an algorithm for the system operation. Even though it was not technically possible to implement this idea, it has become a challenge and inspiration for many modern architects and designers. Most of them intended to create a façade that is "alive and constantly changing like a chameleon skin" (Davies 1981).

Davies's work clearly shows that inspiration from nature can be implemented in (at least) two ways; the bionic innovation could either be inspired by the way the organism functions or it could be modelled on the way that the organism is built, especially at the level of its microstructure. The first method has been widely implemented in systems that are used to maintain stable conditions in buildings (energy conservation and sustainability), while the latter has inspired several innovative material solutions. Observation of animals, especially in extreme weather conditions (e.g., tropical climate or polar zones), provides numerous organic patterns to be used in building systems and components, such as transparent insulation materials (TIM) modelled on the fur of the polar bear. In organisms, the thermal capacity of blood is used in thermoregulation that takes place in the blood vessels. Some animals (e.g., desert animal—fennec fox) take advantage of this mechanism with enhanced emission of excessive heat into their surroundings. Another analogy between the nature and built environment can be seen in the reaction of animal skin on temperature changes and the façade reaction to the solar gain. The use of sunlight for energy production constitutes a technological equivalent of photosynthesis. Many buildings also mimic the natural phenomenon of evaporative cooling. This mechanism is analogous to sweating which takes place in humans and most animals. This process has inspired many technical solutions in modern architecture. After being initially observed in animals and humans, evaporative cooling is effectively used in arid and dry climates to reduce air temperature in buildings (Brzezicki 2002).

The most popular analogy between buildings and nature is the one that compares buildings and warm-blooded organisms. Many similarities could be found between such organisms and the built environment. For example, the process of homeostasis is analogous to the artificial climate regulation in buildings. Another example is the information transfer in the buildings that parallels that found within the nervous system. The maintenance of constant conditions in buildings requires energy and relatively fast reaction times to changing external conditions. Some solutions could be found in nature and directly implemented in some contemporary buildings. Warm-blooded organism analogy contributed to the invention of energy-conserving countercurrent heat exchanger after the similar natural system (veins and arteries

Fig. 1.9 Conceptual façade system modelled on the principle of gaseous exchange in insects. Picture taken during Glasstec Messe 2004, Glass Technology Live Show, Institute of Building Technology and Design, Stuttgart, Prof. S. Behling, Dr. Henning-Braun

intertwined in the limbs) was observed in aquatic birds and mammals. The thermoregulation phenomenon in animals is a continuous inspiration for engineers and architects designing innovative technical solutions and sustainable technologies.

While the condition of the "channelling systems" in an organism is constantly maintained through the biological mechanisms, the same process is difficult and time-consuming in buildings. In this case, the gaseous exchange in insects provided some insights as it is done directly through the exoskeleton through the apertures called spiracles (Fig. 1.9). Further observation of the insect's respiratory systems contributed to the recent invention of decentralized air-condition system that provides direct air exchange with the external conditions, without the use of any channelling systems (Brzezicki 2007). This type of air-conditioning system uses less energy (no energy to pump the air through channels is required) and allows to switch the system on and off depending on the room occupancy. The lack of the channelling permits reducing room storey-height and subsequently the overall height of buildings.

1.4.2 Plants Inspiring Design

Plants have been evolving for approximately 460 million years. Due to constant environmental pressures, they have become extremely well adapted to various climatic conditions (Koch and Barthlott 2009). Due to the immobility of individual plants, they are together an excellent biological material for detecting climate phenomena (López et al. 2017). Living organisms use smart, optimized, and elegant solutions to survive, thanks to continuous evolutionary processes (i.e., selection and mutation). Consequently, plants have developed a tissue with barrier properties after facing a number of survival challenges (e.g., water loss, extreme temperatures, UV and solar radiations, and parasites).

One of the most important improvements in existing air exchange solutions in buildings is the introduction of natural ventilation. The gas exchange by stomatal apparatuses is evidently analogous to processes found in plants. Centralized respiratory and circulatory systems are good models of artificial air-conditioning and centralized heating systems. The latter was not mimicking the natural systems from its conception; still, the natural solutions are often used to illustrate the concept of centralized energy and matter exchange.

Figure 1.10 presents several examples representing adaptation of plants to various environmental constraints. These are often identified as straightforward inspirations for the responsive façade materials or solutions. Adaptations are additionally summarized as possible innovations and challenges that may be implemented in specific façade systems (Table 1.1). The description of environments includes a list of abiotic factors that forced specific plants to attain certain

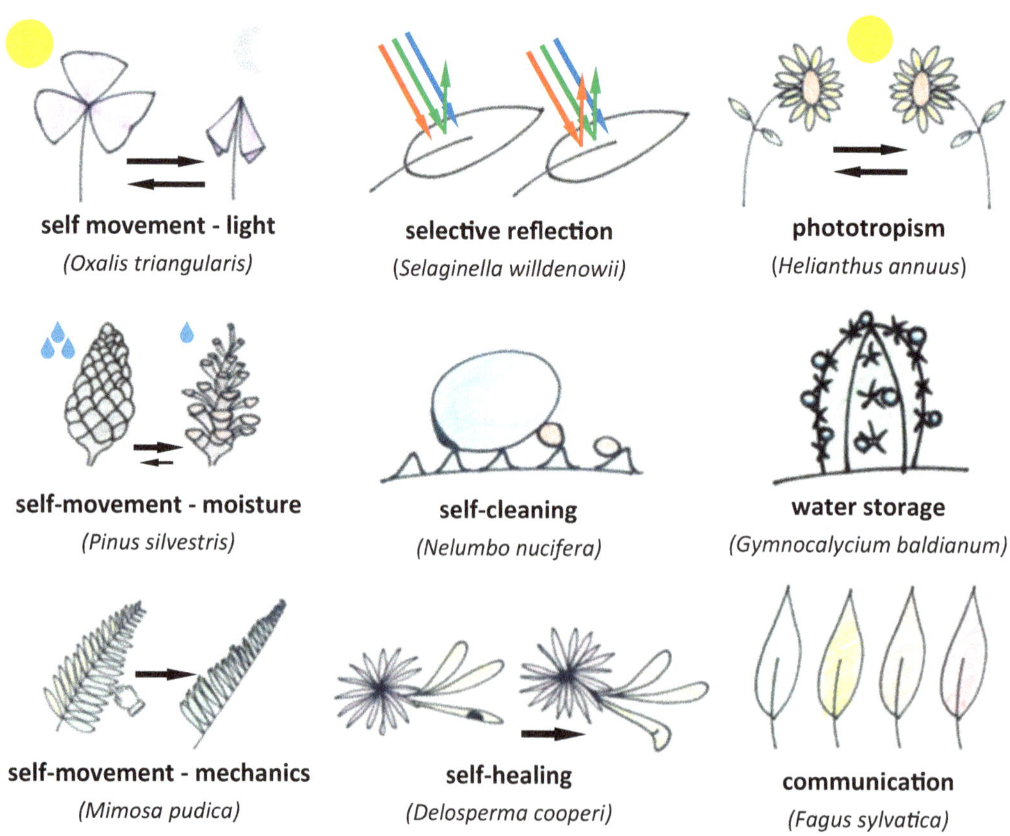

Fig. 1.10 Adaptations of plants and their possible implementation in façade systems

Table 1.1 Biological adaptations of plants and its possible implementation for façade systems

Climate description	Biological adaptation	Façade biomimetic
Desert—Arid (BW)		
Dry Hot Direct sun Strong wind Extreme temperatures Water loss Drought	Thick, small leaf H_2O storing Reduced transpiration Hair and spines Collecting H_2O Thick layer of wax Low H_2O loss Light reflection Long root system H_2O capturing	Reducing evaporation UV protection Shading system Reflective system Filtering system
Prairie—Arid (BS)		
Hot summers Cold winters Strong wind Uncertain rainfall Common drought	Narrow leaves plant shape H_2O storing Low transpiration Protection against animals Thick bark Fire resistance, high regeneration Seed dispersing system Reproduction	Dynamic Opening–closing system Self-healing materials External protection
Rainforest—Tropical (A)		
Hot Wet Uneven solar radiation Heavy rains	Drip tips High water runoff Wax Protection and hydrophobicity Aerial roots High H_2O, CO_2 uptake Low decay Plant morphology H_2O storing	Self-cleaning Filtering Phytoremediation Self-cleaning surfaces External protection
Tundra—Polar (E)		
Short cool summers Long/severe winters Low rainfall Permafrost Solar light variation	Colour of plant Modulated light reflection Plant movement High absorbance of solar radiation Wax and hairs Protection against freeze	Anti-freezing Energy storage Dynamic shading system Modulation of light Transmission Shading system
Temperate Forest—Temperate (C)		
Hot summers Winter bellow 0 °C No problem with H_2O availability Four seasons	Lightweight leaves High photosynthesis Reaction wood High mechanical resistance Nyctinasty, thigmonasty Different response to light and mechanical stress Thick bark High thermal isolation	Reaction to stress Shading and signalling Dynamic energy storage Insulation, shading Self-healing materials Communication

(continued)

Table 1.1 (continued)

Climate description	Biological adaptation	Façade biomimetic
Water—Everywhere ()		
Continuous wet Stable temperature No direct sun Constant H_2O availability Water current, flood	Flexible stem Floating Hydrophobic surface H_2O protection Thin cuticle High CO_2 uptake Floating seeds Reproduction	H_2O storage Stiffness change Hydrophobic surfaces Self-cleaning surfaces Flexibility

adaptations. The following part presents the mechanisms describing how the chemical composition, anatomy, morphology, and behaviour of plants respond to external environment (i.e., protection against excessive wind, drought, water, cold, heat, and light).

Even though naturally evolved biological tissues are based on a relatively simple set of chemical substances (i.e., containing only carbon, nitrogen, oxygen, and hydrogen), plants created a range of materials with outstanding functional properties. In nature, heterogeneity, anisotropy, and hygroscopicity are utilized as response tools with strategies adapted to diverse climatic constraints. Hierarchical structure of natural materials and various properties at different length scales allowed plants to better meet adaptation requirements. Consequently, simple material elements simultaneously act as sensors, actuators, and regulators. Comprehensive analysis and evaluation of plant adaptation strategies (both static strategies and dynamic mechanisms) to their environment in different climate zones are indispensable for transferring concepts from biology to architecture. Thus, unique adaptation solutions can be implemented in new materials that will be used in building envelopes erected in specific climate zones. Integration of length scales together with biological, chemical, and physical concepts for tailoring properties of materials during their preparation should lead to improved design of future smart materials. This optimization process should promote the development of active biomaterials performing as interfaces between outdoor conditions and internal comfort that are able to regulate humidity, temperature, CO_2, and light and also capture and filter pollutants, self-assemble, self-clean, graft, and self-heal. Such materials could be used as responsive building elements and contribute to an improved performance and energy efficiency of building skins.

1.5 Adaptive and Responsive Façades

Unfortunately, standard static façades require constant human attention to regulate microclimate of buildings. Often, even human effort may be insufficient. For example, opening windows to decrease temperature is pointless when the outside temperature is higher than the desired room temperature. For this reason, a new approach for façade design has been proposed—an adaptive façade. The term "adaptive façade" covers the façade systems capable to change its function, shape, and behaviour in response to the fluctuating external conditions. In buildings, adaptive façade systems (called also "responsive" and "dynamic"), may assure controllable insulation, radiant heat exchange, daylighting, solar shading, humidity control, ventilation, and energy harvesting (Loonen et al. 2015). Façade adaptivity (a self-regulation of certain façade's properties) can be manifested in several manners: by the physical change of the façade shape (so-called kinetic façades), by the active control of the energy flow (e.g., by opening and closing windows, retracting sun shades, or operating fans), or by the energy harvesting (e.g., by solar collectors or photovoltaic panels) (Perino and Serra 2015). The idea of adaptive façade is well integrated with EU idea of near-zero energy building policy, since adaptive façades have a positive impact on the quality of the indoor environment due to significant reductions in building energy use and CO_2 emissions (Loonen et al. 2015).

Mixed-mode systems have been frequently fulfilling comfort needs of occupants, but they have been rarely optimized from the perspective of energy efficiency and carbon emission reduction. This is the reason why adaptive façades systems were first theorized and then implemented into the real buildings. The first concept of an adaptive façade included a multi-layered, multifunctional barrier that would autonomously adapt to changes in the external environment, similar to the human or animal skin (Davies 1981). Throughout the years, this idea has been a constant inspiration for the architects to design new façade systems. Adaptive façade assumes a certain level of façade autonomy. Adaptation takes place automatically, without the need of user attention. This decreases building energy use and simultaneously improves internal microclimatic conditions. Both objectives can be reached while still considering occupant needs, as it is always possible to override system settings manually and thus influence internal microclimatic conditions according to personal preferences.

In general, the idea of façade adaptivity assumes that the systems created would be more energy efficient and environmentally friendly when compared to the system that is static and unable to change any of its properties. It is relatively simple to achieve good energy efficiency in a moderate climate when the outside air temperature is suitable to act as a microclimate regulating medium. In this case, the outer air can effectively cool down the room while taking the heat by the means of simple air exchange, implemented usually as displacement ventilation. Façade systems could provide the energy to the grid by generating, storing, and distributing energy. Such systems can significantly contribute to the global decarbonization goals that foresee 80% reduction of greenhouse gas emissions by the year 2050. However, façade solutions with all the functions listed above currently do not exist. The most up-to-date prototypes can handle a limited number of operations simultaneously (e.g., daylight or/and ventilation control). Fortunately, several innovative solutions are either already emerging in the market or are at the advanced conceptual stage.

1.5.1 Type of Control

In adaptive façades, there are two basic control types: extrinsic and intrinsic. Extrinsic (meaning "directed from outside") type is reflected in three phases of operation: collecting the information about the environment (detecting, sensing); processing the information (computing according to certain algorithms), and taking the physical actions (e.g., actuating, folding, rolling, expanding). These three stages are usually executed by the system based on the propagation of electronic signals (e.g., when a detector senses high levels of illuminance). First, the actuator's displacement is calculated by the integrated computer system, and later, the actuator rotates the louvers to the proper position. This process could be also executed through other means, such as the pneumatic mechanism.

The intrinsic control type relies on the inherent properties of the used components/elements and their self-behaviour/adjustment according to the external conditions. The components/elements change size, shape, volume, phase, or colour when subjected to various environmental triggers and can thus contribute to the improved façade performance by helping with the regulation of certain façade features. This type of control is self-autonomous, as illustrated by, for example, the shape changes of the bimetallic element when under the influence of temperature, which cannot be externally regulated. Bimetallic element returns to its original shape only after the initial influencing factor ceases to act (e.g., the temperature drops after sunset).

1.5.2 Practical Implementation

In practice, integrated façade solutions that would fulfil all the initially postulated features do not yet occur. Selected functionalities are being implemented individually, as pilot sub-systems of larger façade schemes. Daylight regulation was the first automated sub-system. Mechanically or pneumatically controlled blinds, sunshades, or apertures require relatively simple operation schemes, with a photosensor as the light condition detector. The control system is relatively simple and could be intrinsic (meaning autoreactive), as the shading response might be independent of any other functional systems of the building. The shading systems can, however, take various forms: from classic lamellar blinds rotating around their own axis, through rolled shutters, to complex geometrical multifaceted kinetic systems that—like paper origami toys—fold in or unfold when reacting to changes in the daylight level. An Oval Cologne Office in Cologne (arch. Sauerbruch Hutton Architekten, 2010) could be an example of a building aimed to regulate daylight levels in office interiors (Brzezicki 2018)—Fig. 1.11

Daylight regulation can also be executed by systems without any mechanical parts with the advantage of the reliability and low operating costs. Such solutions are based on electrochemical reactions in fine layers of semiconducting materials. This includes

Fig. 1.11 An oval cologne office in cologne (arch. Sauerbruch Hutton Architekten, 2010)

electrochromic coatings that recently gained a lot of attention with many commercialized solutions available under various brand names (e.g., SageGlass). The main obstacles to a wide use of these technologies are relatively high cost and elongated dimming time (7–12 min are necessary to change the material phase). Alternative systems that react immediately and are based on liquid crystals (e.g., from Merck) recently entered the market and are under pilot installations.

The adaptive regulation of other façade parameters causes more technical and practical problems. Materials with variable heat transfer coefficient (so-called regulated insulation) are recently in the prototype phase. In most cases, the insulation action is executed by regulating convection—by creating adjacent air buffers and skilfully controlling the air circulation between them (Pflug et al. 2018). It is also possible to regulate the thermal capacity of the façade so that it can accumulate energy during periods of increased heat load. The latent internal energy for the physical phase change is used. This energy affects the melting/solidification of specific materials without the increase of the insulation layer temperature. At present, certain types of paraffin can accumulate heat during the daytime in a process of melting (phase change from the solid to liquid state) to later return that heat at the cooler night-time by solidification (phase change from the liquid to solid state).

References

Bribián IZ, Capilla AV, Usón AA (2011) Life cycle assessment of building materials: comparative analysis of energy and environmental impacts and evaluation of the eco-efficiency improvement potential. Build Environ 46(5):1133–1140

Brzezicki M (2002) Wzorce organiczne w projektowaniu fasad ekperymentanych (Organic patterns in the design of experimental façades). In: Nyka L (ed) Architektura współczesna wobec natury. Faculty of Architecture, Gdansk University of Technology, Gdańsk, pp 86–94

Brzezicki M (2007) I-modul Fassade—przełom w regulacji mikroklimatu budynku. Świat Szkła 2:20–22 (in Polish)
Brzezicki M (2018) Inteligentne fasady. Architektura Murator, suplement. In: Warsztat Architekta, vol 2, pp 53–58 (in Polish)
Capeluto G, Ochoa CE (2017) Intelligent envelopes for high-performance buildings design and strategy. Springer International Publishing Switzerland
Davies M (1981) A wall for all seasons. RIBA J 88(2):55–57
Gruber P, Gosztonyi S (2010) Skin in architecture: towards bioinspired façades. WIT Trans Ecol Environ 138:503–516
Gruber P (2011) Biomimetics in architecture. In: Gruber P, Bruckner D, Hellmich C et al (eds) Biomimetic–materials, structures and processes examples ideas and case studies. Springer, Berlin
Herzog T, Krippner R, Lang W (2004) Façade construction manual. Birkhäuser, Basel
Herzog T, Natterer J, Schweitzer R et al (2012) Timber construction manual. Birkhäuser, Basel
Koch E, Barthlott W (2009) Superhydrophobic and superhydrophilic plant surfaces: an inspiration for biomimetic materials. Phil Trans R Soc A 367:1487–1509
Knaack U, Klein T, Bilow M et al (2007) Façades: the principles of construction. Birkhäuser, Basel
Lechner N (2008) Heating cooling and lighting: sustainable design methods for architects. Wiley, New Jersey
Loonen RCGM, Rico-Martinez JM, Favoino F et al (2015) Design for façade adaptability—towards a unified and systematic characterization. In: Proceedings of 10th conference on advanced building skins, Bern
López M, Rubio R, Martín S et al (2017) How plants inspire façades. From plants to architecture: Biomimetic principles for the development of adaptive architectural envelopes. Renew Sustain Energy Rev 67:692–703
Lovel J (2013) Building envelopes: an integrated approach. Princeton Architectural Press, New York
McDade JE, Shepard CC, Fraser DW et al (1997) Legionnaires' disease: isolation of a bacterium and demonstration of its role in other respiratory disease. N Engl J Med 297(22):1197–1203
Pawlyn M (2016) Biomimicry in architecture. Riba Publishing, Newcastle
Perino M, Serra V (2015) Switching from static to adaptable and dynamic building envelopes: a paradigm shift for the energy efficiency in buildings. J Facade Des Eng 3(2):143–163
Pflug T, Nestle NE, Kuhn T et al (2018) Modeling of facade elements with switchable u-value. Energy & Buildings 164:1–13
Pohl G, Nachtigall W (2015) Biomimetics for architecture and design. Springer International Publishing Switzerland
Samuel DGL, Dharmasastha K, Nagendra SMS et al (2017) Thermal comfort in traditional buildings composed of local and modern construction materials. Int J Sustain Built Environ 6 (2):463–475
Siegesmund S, Snethlage R (2014) Stone in architecture properties, durability. Springer, Berlin
Watts A (2014) Modern construction envelopes. Birkhauser, Wienna

Chapter 2
Biomaterials for Building Skins

Timber, despite being the world's oldest construction material, is now the most modern.

Alex de Rijke

Abstract Bio-based materials are considered a promising resource for buildings in the twenty-first century due to their sustainability and versatility. They can be produced locally, with minimum transportation costs and in an ecological manner. This chapter describes the potential of biomaterials for use in façades. It presents several examples of natural resources, including innovative alternative materials that are suitable for implementation as a building skin. Novel products resulting from material modifications and functionalization are presented, including a brief discussion on their environmental impacts. Alternative strategies for optimal biomaterials' recycling, reuse, and other end-of-life strategies are presented and supported with case study examples.

2.1 Why Build with Biomaterials?

The current trend for constructing sustainable buildings and increasing environmental awareness is reviving bioarchitecture as an alternative to other construction techniques. The unique properties and the natural beauty of bio-based materials make them desired in various applications, including construction and interior/exterior design, among others. The main advantage of biomaterials is the low environmental impact due to their renewability and cascade use. Only low amounts of energy are needed to manufacture wood. The production of wood as a building material involves only about 10% of the energy consumption required to produce equivalent amount of steel (Odeen 1985). Moreover, it can be processed using simple tools. Biomaterials enable prefabrication and fast installation. Due to a favourable weight-to-load-bearing capacity ratio, they enable erection of multi-storey structures while enabling considerable design freedom. The low thermal conductivity of timber increases its applicability in the façade interface between the inside and the outside (Tapparo 2017).

A. Sandak et al., *Bio-based Building Skin*, Environmental Footprints and Eco-design of Products and Processes, https://doi.org/10.1007/978-981-13-3747-5_2

2.1.1 Unique Characteristics of Biomaterials

Wood is highly recyclable, and several reuse options make it an excellent material for the currently desired cascade use. Wood and other bio-based building material products have the advantage of a significantly lower carbon footprint than steel, glass, or concrete (Tellnes et al. 2017). Since trees absorb CO_2 from the atmosphere and store carbon in the wood tissue, wood generates lower environmental impact in comparison with other building materials. Consequently, biomaterials have become recognized as an attractive alternative to several traditional building solutions, making biomaterials "building materials of the twenty-first century" and "timber a new concrete". Since biomaterials can efficiently sequester carbon, they are counterbalancing emissions from other materials. However, compared with traditional building materials, biomaterials possess some properties that are less understood and remain difficult to control. Natural fibres, for example, are capable of binding the amount of moisture equivalent to between 5 and 40% of their dry weight, depending on the ambient air conditions. These fibres can then act as buffers or humidity/water absorbers within certain building structures. However, the ability to bind moisture influences the hygro-thermal stability of components made of hygroscopic materials. Hygro-thermal stability is an important constraint in certain applications, such as thermal insulation, hydrocivil engineering, cladding, and decking. Hygroscopic properties of any bio-based material are the main reason for shrinkage and swelling, and, consequently, for dimensional distortions. The interrelation between the relative humidity (RH) and equilibrium moisture content (EMC) of a material is usually presented as a sorption isotherm diagram (Willems 2014; Brischke 2017). The moisture content decreases during the drying, until the free water completely evaporates, but the entire amount of bounded water remains. This state is defined as a fibre saturation point (FSP).

Controlling biomaterial moisture content is crucial to avoid decay process. The minimum moisture content stimulating fungal growth is estimated to be from 22 to 24%, depending on the material type. For this reason, building experts recommend 19% as the maximum limit of the moisture content in untreated wood to assure its safe service. The best practice design uses the "4 Ds" to limit the amount of moisture intrusion. The first line of defence is (1) deflection, that deflects water away from the structure. A small amount of water that can pass the cladding should exit the wall via (2) drainage path. All the remaining water should be able to easily (3) dry. Finally, it is recommended to use (4) durable materials, such as naturally decay-resistant species or preserved/modified wood. Excessive moisture in wood structures makes these more susceptible to insect attack. There are several guidelines which include basic protection practices (e.g., in case of termites: maintaining structure dry, applying chemical termiticide to the soil, using barriers and traps). These should be supported by maintenance practices that consist of regular inspections (Reinprecht 2016). With a proper design, construction, and maintenance, timber buildings provide a service that is at least equivalent to other building types (Foliente 2000).

Bio-based materials have the disadvantage of being combustible; thus, they are perceived as less safe than steel and masonry. Combustibility limits their use as a building material due to restrictions in building regulations in most countries, especially in taller and larger buildings (Buchanan et al. 2014). The recently released document "Fire Safety in Timber Buildings—Technical Guideline for Europe" presents the background and design methods for designing timber buildings to assure comparable levels of fire safety to buildings made of other materials (Östman 2010). The most recent report "Fire Safety Challenges of Tall Wood Buildings" by Gerard et al. (2013) is not limited to European contexts but includes case studies of modern timber buildings from around the world. Modern building codes should move towards performance-based design for fire safety, that is, towards designing the building to achieve a target level of performance rather than simply meeting the requirements of a prescriptive building code. The European Construction Products Directive (CPD) has introduced five essential requirements of fire safety. The structures must be designed and built in a way that the load-bearing capacity, in the case of a fire event, will assure structure to remain intact for a specific period of time. The generation and spread of fire and smoke in the building, as well as the spread of fire to neighbouring structures, must be limited. In the event of fire, occupants may either leave the building on their own or be rescued by other means, where the overall safety of rescue teams must be taken into consideration.

Automatic fire sprinkler systems, being the most effective way of improving the fire safety, are especially recommended in tall buildings. Another approach to improve fire resistance is encapsulation (complete, limited, or layered) of timber elements with non-combustible materials. The use of fire-resistant and fire-retardant coatings/treatments is an additional method to improve fire performance of bio-based materials. Fire resistance is a property of a material to withstand fire or provide protection from it. Fire retardants reduce the amount of heat released during the initial stages of a fire and reduce the number of flammable volatiles released during the subsequent fire stages. Fire retardants can contribute to diminishing fire propagation by protecting the surface through insulation layers, changing the pathway of pyrolysis, slowing down ignition and decreasing burning temperatures by changing the thermal properties of the product, reducing combustion by diluting pyrolysis gases, and reducing combustion by inhibiting the chain reactions of burning. In practice, most retardant systems combine diverse mechanisms and, in consequence, increase their own overall efficiency (Russel et al. 2007). On the other hand, fire itself can be used as a protective treatment. The Shou Sugi Ban technique developed in Japan to protect the external cladding made of cedar involves charring a wood surface (Fig. 2.1). Even though this technique was established in the eighteenth century, it is considered a highly interesting treatment for contemporary exterior and indoor spaces (Fortini 2017).

An important advantage of using biomaterials is their naturalness and other assets compatible with the human physiological deactivation. Modern trends in building design tend to move beyond the simple optimization of basic environmental characteristics (such as air temperature and humidity), to more holistic approaches that

Fig. 2.1 Shou Sugi Ban surface treatment of building façades

have a health-supporting role. Biophilic design rediscovers an ancient practice, where humans were a part of a wider ecosystem integrating natural elements into built structures. In architecture, biophilic design is a sustainable design strategy that reconnects people with the natural environment. It focuses more on the human well-being than to green building principles that emphasize the responsibility to the environment and efficient use of sustainable resources. Bio-based building materials can favourably affect occupants of the built environment. Clear benefits of natural materials on human health, childhood development, health care, learning, work efficiency, and productiveness were reported (Kellert et al. 2008; Kotradyová 2013; Kotradyová and Kalináková 2014). It should be mentioned, however, that industrial transformation, post-processing, and modifications may highly affect human perception of material "naturalness" (Burnard et al. 2017).

2.1.2 Sustainability of Natural Resources

Wood and other bio-based materials can be produced locally, with minimum transportation costs and in an ecological manner. Forest certification is a recent approach ensuring that forest products are produced sustainably. Certification aims to improve the image of timber producers and explains their involvement to sustainable forest management. Global regulations requiring the exercise of "due diligence" and "risk assessment" are becoming more common. The European Union Timber Regulation (EUTR) prohibits placing on the EU market wood harvested in contravention of the applicable legislation in the country of origin, as well as wood products derived from it. This applies to both imported and domestically produced timber and timber products, such as solid wood products, flooring, plywood, pulp, and paper (ECE/TIM/SP/33 2013). The most common certification marks used in the forest industry are presented in Fig. 2.2.

Fig. 2.2 Forest certification marks: **a** "Conformité Européene", **b** Forest Stewardship Council, and **c** programme for the endorsement of forest certification

Frequently used Conformité Européene (CE) indicates that the product conforms to all applicable European legislation related to safety, health, energy efficiency, and environmental concerns. Forest Stewardship Council (FSC) is an international organization founded in Europe in 1993 that provides a system for voluntary accreditation and independent third-party certification. The system permits certificate holders to mark their products and services along the production chain. It confirms that wood products are coming from well-managed forests that provide environmental, social, and economic benefits. The Programme for the Endorsement of Forest Certification (PEFC) is an international non-profit, non-governmental organization endorsing sustainable forest management (SFM). It ensures that timber and non-timber forest products are produced by respecting ecological, social, and ethical standards. PEFC is certified over 230 million hectares of forests in 28 countries (ECE/TIM/SP/33 2013).

2.2 Natural Resources

2.2.1 Timber

Wood is one of the earliest building materials. Its availability, relatively low maintenance cost, and easy processing make it a prevalent construction material in both interior and exterior applications. Various types of load-bearing structures, as well as complementary construction components, such as cladding, decking, doors, and windows were, and still are, frequently made of wood. The performance and strength of wood used in structural applications are influenced by its physical properties, such as density, mechanical resistance, sorption and permeability, dimensional stability, thermal conductivity, acoustic and electric properties, natural durability, and chemical resistance (Mazela and Popescu 2017). In addition, appearance, smell, morphology, roughness, smoothness, and specific surface area, among others, are important material characteristics influencing perception of materials and interaction with them.

All renewable biomaterials are highly naturally variable. This is expressed in their intrinsic characteristics (Fig. 2.3). Wood, for example, exists in a wide range of colours, patterns, and gloss levels, depending on the species and finishing

Fig. 2.3 Variety of colours and textures of biomaterials suitable for use in building façades

applied. Tannins, pigments, and resins make wood more colourful (ranging from white in case of aspen to black in case of ebony). Many hardwood species (maple, oak, beech, elm) have a characteristic lustre that increases their gloss. The wood anatomy—yearly rings, rays, and fibres—provides unique texture that can be additionally highlighted by protective treatments (e.g., oil or wax finishing). The colour of a material (especially if not protected) changes with time and becomes darker (in case of ageing) or greyish (in case of weathering). However, timber does not only vary in terms of appearance. It mainly varies in physical (hygroscopic properties, density, shrinkage-swelling, as well as sound transmission, electrical, and thermal conductivity) and mechanical properties (strength, toughness, hardness, elasticity, plasticity, brittleness, wear resistance).

The density is defined as a ratio of the mass to the volume. Timber species are classified into six classes regarding their density, with "very heavy wood" class consisting of wood dense more than 800 kg/m^3 (represented by hornbeam, yew, or ebony). The opposite, "very light wood" class contains species with the density lower than 400 kg/m^3 (that includes poplar, white pine, and balsa). Different wood species differ in their natural resistance to biological attacks, including resistance to both wood-decaying fungi and wood-destroying insects. In most wood species, the sapwood (the living part of the standing tree involved in the growth of the plant) is not resistant to biological attacks, while the durability of heartwood (central part of the steam not involved in the sap flow) is very variable. The standard EN 350 (2016) "Durability of wood and wood-based products—Testing and classification of the durability to biological agents of wood and wood-based materials" provides guidance in determining and classifying the durability of wood and wood-based materials against biological wood-destroying agents. The following tables provide an overview of the wood classification into diverse durability classes, considering

Table 2.1 Durability classes of wood-based materials: resistance against attack by decay fungi

Durability class	Description
DC 1	Very durable
DC 2	Durable
DC 3	Moderately durable
DC 4	Slightly durable
DC 5	Not durable

Table 2.2 Durability classes of wood-based materials: resistance against attack by termites, marine organisms, and wood-boring beetles

Durability class	Description
DC D	Durable
DC M	Moderately durable
DC S	Not durable

Note In case of wood-boring beetles, there are just two durability classes, DC D and DC C

resistance against wood-decaying fungi (Table 2.1), beetles boring dry wood, termites, and marine organisms capable of attacking wood in service (Table 2.2).

In the building context, mechanical properties are the most important characteristics of wood as a structural material. These encompass wood's ability to resist distortions and deformations (elastic properties) or failure (strength properties). Mechanical properties of timber and engineered wood products are influenced by environmental factors. Changes in moisture, temperature, pH, decay, fire, and UV radiation can significantly change strength properties (Mazela and Popescu 2017).

2.2.2 Non-wood Biomaterials

The use of non-wood materials in the built environment is an area of growing importance. These are successfully used in roofing, wall constructions, wall cladding, insulation, composites, and chemicals used in the built environment. Renewable materials are already utilized on a significant scale worldwide, with an estimated 71 million tons of crop-derived industrial materials produced annually (Hodsman et al. 2005). Fibre crops, such as hemp and flax, are used in textiles, paper, composites, construction packaging, filters, and insulation. The key market sectors for European fibres in the built environment are wood-based panels, fibre-reinforced composites, fibre–cement composites, and insulation products.

Flax, hemp, cereal straw, Miscanthus, sisal, jute, and kenaf have been used in Europe for producing panels (2 million tons in 2010), fibre-reinforced composites (0.25 million tons), and insulation products (no data available) (Hodsman et al. 2005). The technology for manufacturing flaxboards (panels in which shives from the stalk of the flax plant are bonded together with a synthetic resin) is slightly different than the technology in mass-produced particleboards. The flax shives are in a form of particles already after pre-processing and therefore chipping operation is not necessary. Flaxboards are used in dry environments in door cores, fire-check doors, and partitions. Flax fibres are also used in insulation products and as additives in cementitious composites (Réh and Barbu 2017a).

Hemp is another widespread plant gaining interest in the building sector. Its insulation capacity, mechanical resistance of the fibres, and low density (110 kg/m^3) are highly relevant for the construction industry. Hemp is an essential ingredient of environmentally friendly building materials. It is used in the manufacturing of reinforced composites, insulation, hempcrete, and lime–hemp mixtures (Réh and Barbu 2017b).

The traditional use of straw includes roofing, bales in load-bearing walls, and substrates for plasters. Straw is widely available and is considered an affordable building material. Traditional buildings using straw in roofing are present all over the world (Fig. 2.4). When implemented appropriately and protected from the moisture uptake, straw is a long-lasting, durable, load-bearing, and insulating material (Walker et al. 2017).

Reed is a traditional material used mainly in roof construction in various parts of the world. When properly used, it does not require maintenance for relatively long periods of time, typically lasting more than 50 years. However, in case of poor raw material quality and incorrect installation detailing, the maintenance-free period may decrease significantly, in extreme cases to less than 10 years. Previous studies showed that the properties of reed are highly related to its origin as well as to harvesting methods and periods (Greef and Brischke 2017).

Fig. 2.4 Use of straw in roof coverage

Grass is one of the most widespread and available renewable materials. Hay deriving from meadow plants was often stored in attics in order to serve as a feedstock for domestic animals and provide thermal insulation in rural buildings. Currently, several laboratory trials in manufacturing composites and insulation products on the basis of grass are ongoing. It is thus expected that grass-based products will reach the market and gain increased interest in the near future (Teppand 2017).

The use of wool and other animal hair in the building sector is currently limited to thermal and acoustical insulation. Thermal attributes, prevention of vapour condensation and mitigation of global warming, that are related to such materials, make them useful in both traditional and modern constructions. Wool has a moisture buffering effect indoors, where fibres capable of absorbing moisture in wet conditions may release it when the ambient relative humidity is low (Mansour and Ormondroyd 2017).

Bamboo and rattan are abundant, renewable, recyclable, and biodegradable materials available in high quantities, especially in Asia. Bamboo is one of the fastest growing plants in the world, simultaneously having low density and high mechanical strength and stiffness. The natural durability of bamboo is relatively low but varies among different species and provenances. Both bamboo and rattan fibres are widely used in manufacturing composites. In this case, fibres can be used in the native form as well as modified chemically or thermally to enhance the properties of the composite (Knapic et al. 2017).

The latest trends in the development of alternative building materials have resulted in the development of mycelium-based composite materials for the use in design and architecture. The natural ability of saprophytic fungi to bind and digest ligno-cellulose is used to manufacture packaging, textile, edible films as well as building and insulation materials (Attias et al. 2017). Commercial composite board (Myco-board) can be utilized similarly as medium density fibreboard (MDF) with the significant advantage of not containing formaldehyde. Recently designed Hy-Fi Mushroom Tower pavilion at the Museum of Modern Art in New York by David Benjamin is the first large-scale structure to use mushroom brick technology (Fig. 2.5). The 13-m-tall tower was created in order to provide a new definition of sustainability. The lightweight bricks being an innovative combination of corn stalk waste and living mushrooms returned to the earth through composting at the end of the structure's lifecycle. In contrast to typical temporary architecture, Hy-Fi was designed to "appear as much as to disappear".

The Cuerden Valley Park Trust (Fig. 2.6) is a superb example of a building that was erected using local natural resources. It was designed by Straw Works according to the Living Building Challenge standard. It has a hybrid load-bearing straw and a timber frame built by volunteers from straw, timber, cedar shingles, lime plaster, sheep wool, and hemp. The Trust building is a visitor centre enabling close connection with the nature.

Fig. 2.5 Hy-Fi Mushroom Tower pavilion at the museum of modern art in New York Photograph courtesy of Amy Barkow

Fig. 2.6 Visitors' centre in Cuerden Valley Park Trust during the construction (November 2017)

2.3 Modification and Functionalization of Biomaterials

Recent advances in biomaterials research have delivered several solutions for the construction sector. Currently industrialized engineered wood products, such as glue-laminated timber beams (glulam) or cross-laminated timber panels (X-lam or CLT), allow using wood for erecting long-span and/or multi-storey buildings. Biomaterials can act as buffers or sinks for water within certain building structures and provide comfort to occupants of the built environment. On the other hand, their ability to bind moisture influences hygro-thermal stability, which can be an important constraint in certain applications, for example, in thermal insulation, hydro civil engineering, cladding, and decking (Jones and Mundy 2014).

To broaden their applicability, biomaterials need to improve in several of their properties, such as dimensional stability, thermal stability, fire resistance, biotic and abiotic degradation resistance, and mechanical properties. This brings new solutions to the market that assures expected properties and functionality over elongated service lives and reduces the risk of product failure. These include novel bio-based composite materials (e.g., fibreboards, particleboards), as well as more effective and environmentally friendly protective treatments, such as thermal treatments, densification, impregnation, and chemical modifications. The same revolutionary progress is observed in surface treatments, including innovative coatings, impregnations, or integration of nanotechnology developments to protect biomaterials. The latest trends are driven by the biomimicry approach of capturing and exploiting properties that have evolved in the nature. All above-mentioned treatments lead to changes of natural properties of wood or other natural resources.

Biomaterials, originate from nature and thus possess great variability in properties that are manifested in different durability/use classes, density, or appearance. The variation is observed between species, trees, and even within the single tree itself. While possessing several advantages, such as aesthetical appeal and positive weight-to-load-bearing capacity ratio, unprotected wood suffers when exposed to environmental conditions and changes its dimensions and appearance (e.g., colour, roughness, glossiness). Wood modification processes enhance desired properties by applying chemical, biological, or physical agents (Hill 2006). Properties of wood are determined by its chemical composition; therefore, most (but not all) modification processes target material changes at molecular levels.

2.3.1 Thermal Modification

Thermal treatment is an effective way to improve properties of biomaterials. The permanent modification of chemical composition of constitutive polymers is achieved through exposure to elevated temperatures with reduced oxygen availability. Thermally modified timber (TMT) is a result of such a treatment (in the case of wood), where the modification process is defined in CEN/TS 15679 (2007). The usual range of the treatment temperatures is between 160 and 230 °C, depending on the intended treatment intensity. Several modifications of wood chemical components occur due to the thermal treatment. Hemicelluloses are deacetylated, depolymerized, and dehydrated to aldehydes. Cellulose microfibrils are less affected by the thermal treatment due to the crystalline structures that are more resistant to the thermal degradation. However, the overall crystallinity level increases because of the amorphous part degradation. Lignin is softened and cross-linked with other cell wall components, which results in the increase of the apparent lignin content (Esteves and Pereira 2009). Thermally modified wood possesses superior durability against decay and weathering, enhanced dimensional stability, constant colour within the bulk, reduced thermal conductivity, lowered equilibrium moisture content, and increased hydrophobicity. These properties partly result from the reduced accessibility of hydroxyl groups to water molecules. Thermal treatment processes can change non-durable softwood species (class 5 according to European Standard EN 350 (2016)) into the superior durability class 1 (Navi and Sandberg 2012). There are several commercially available technologies differing in terms of treatment conditions (e.g., temperature and duration, steam presence or absence, atmosphere type, use of catalysts, closed versus open system) (Gérardin 2016).

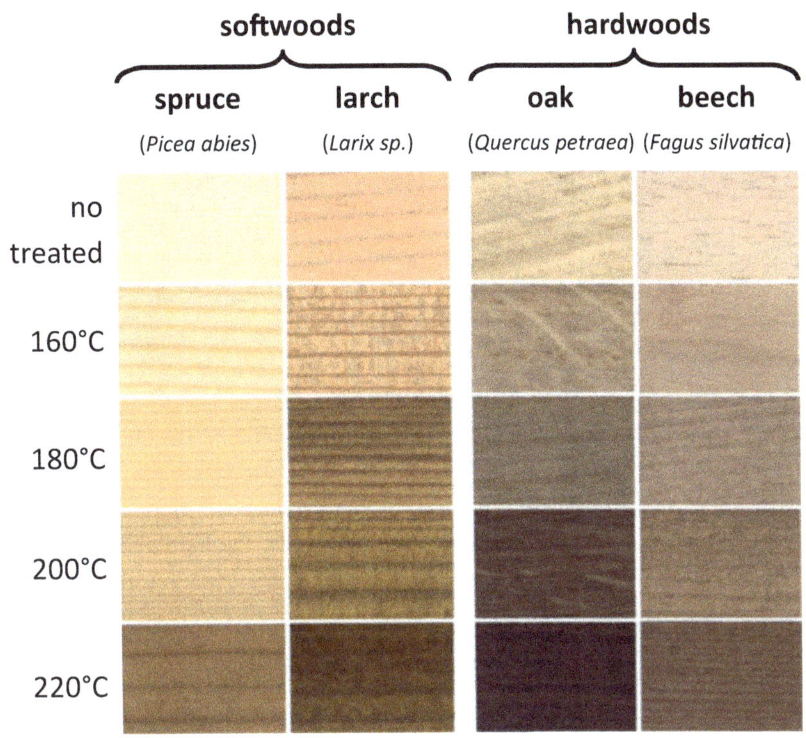

Fig. 2.7 Effect of the thermal modification on the appearance of wood

Properties of TMW depend on the settings of the treatment process and wood species, among other factors. In general, thermally modified wood is not suitable for load-bearing applications. Its dimensional stability is improved with a more hydrophobic surface. This has an effect on the surface finishing procedures and the coating performance. The colour becomes darker as a result of thermal modification although the discoloration is not stable when used in the exterior. The surface of thermally modified wood becomes grey/silver after a short weathering. Even though the decay resistance of TMT is improved, it is not recommended to use it in contact with the ground. The appearance of selected wood species that were modified thermally at different temperatures is presented in Fig. 2.7.

2.3.2 Chemical Modification

Chemical modification is the reaction of a chemical agent with wood chemical components resulting in formation of covalent bonds (Hill 2006). Acetylation is the most established treatment, where acetic anhydride reacts with hydroxyl groups of cell wall polymers by forming ester bonds. The reaction replaces hydroxyl groups with acetyl groups and yields acetic acid as a by-product. Acetylation improves UV resistance and reduces surface erosion by 50%, which is important when using wood as a façade material (Rowell 2006). The mechanical strength properties of

Fig. 2.8 Acetylated wood as a material for a building façade. Photograph courtesy of Accsys

acetylated wood are not significantly different than in untreated wood; however, its durability is substantially improved. An example of the use of acetylated wood is presented in Fig. 2.8. Suitability of other reagents as an alternative to the acetic anhydride was investigated for wood modification. Unfortunately, most of those technologies were never implemented as industrial solutions were not commercialized.

2.3.3 *Impregnation*

Impregnation process leads to locking selected chemicals within the wood cell wall. The cell wall should be in a swollen state to ensure accessibility of the impregnate. The treatment is considered effective when chemical substance used for impregnation is not leachable in-service conditions. Several substances are currently used for impregnation, such as resins (UF, PF, MF, MMF, DMDHEU), furfuryl alcohol, and inorganic silanes, among the others. Some of the processes, like furfurylation (Kebony®) or DMDHEU treatment (Belmadure®), are commercialized, and their products are available on the market. When considering the large variation of the impregnation methods, it becomes clear that the performance of impregnated wood can vary significantly. However, in general, these treatments reduce swelling and shrinking, improve dimensional stability, and increase resistance to biotic degradation. Figure 2.9 presents the use of furfurylated wood in an innovative architectural sculpture inspired by nature.

Fig. 2.9 Furfurylated wood as a material for a building façade. Photograph courtesy of Kebony

2.3.4 *Surface Treatment*

The above-mentioned treatments are related to modifications occurring within the whole volume of the material bulk. In contrast, several techniques affect only the properties of the surface without interfering with the interior of the material. The changes of the surface functionalities affected by the exterior treatments include UV stabilization (e.g., surface acetylation), increase of hydrophobicity (e.g., reaction with silicone polymers), or improvement of the adhesion (e.g., enzymatic treatment, plasma discharge). In addition, additional processes can be applied to improve biomaterial surface resistance against biotic and abiotic factors, such as surface densification (Rautkari et al. 2010) or surface carbonization (e.g., Shou Sugi Ban).

Façade surface finishing by diverse coatings, waxes, oils, or stains is the most common treatment of the surface that highly influences its service life performance. The systematic comparison between different finishing technologies is presented in Fig. 2.10. The resistance of the surface against deterioration in service highly depends on the finishing product quality (i.e., chemical formulation), surface preparation (e.g., oxidation stage, roughness, wettability, surface free energy), and the application procedure (e.g., industrial coating, immersing, brushing, or spraying). A high variety of commercially available products for surface finishing can produce various differences in appearance, including variations in colour, transparency, and gloss. An example of appearance changes in wood surfaces when treated with different surface coatings is presented in Fig. 2.10. A proper use of the surface finishing technologies may highly contribute to the aesthetical attractiveness of the structures as well as to the changes in appearance along the service life of the façade. The cost of the finish, including proper surface pre- and post-treatment, may be substantial. However, it may be sensible, from a financial point of view, to rise the initial cost of the façade by increasing the thickness of the coating layer, as this may significantly extend the time of maintenance-free use. It has been reported that

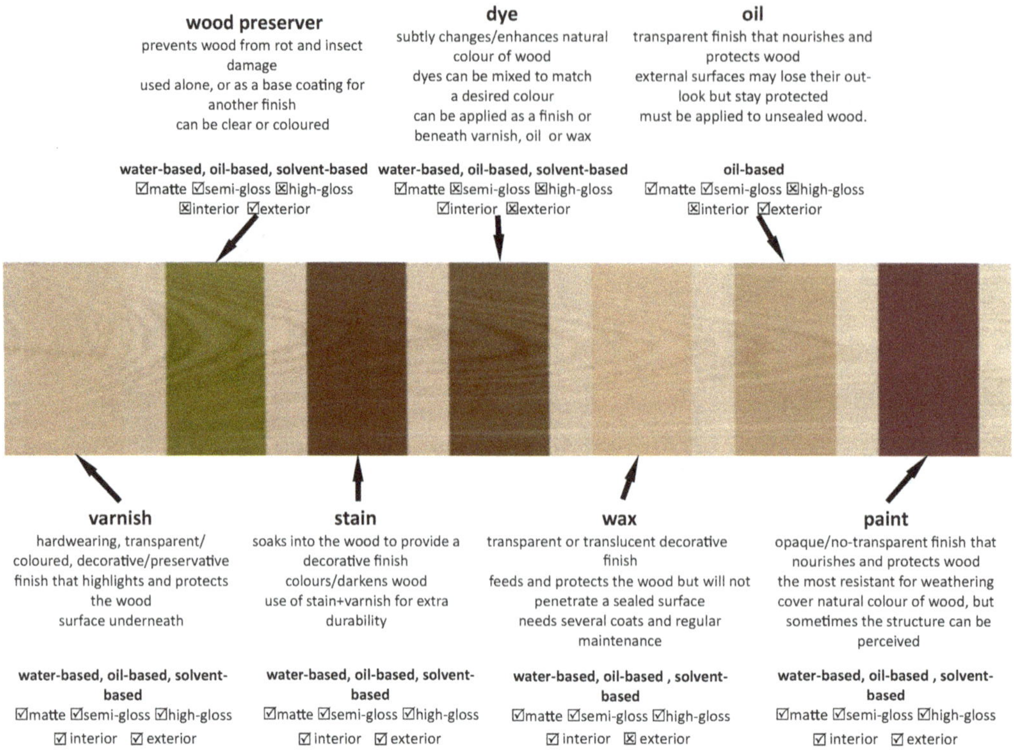

Fig. 2.10 Appearance of the wood surface after diverse coating solutions

the increase of the coating layer thickness from 30 to 50 μm extends the time of the surface resistance against cracking, and the service life period, by a factor of 1.2 (http://www.servowood.eu).

2.3.5 Hybrid Processes

Bulk and surface modification processes may affect one or more functionalities of the biomaterial used in a building façade. Although each modification process improves certain material properties on its own, this positive effect can be multiplied by merging two or more modification processes. Such an approach is a "hybrid process" of biomaterial modification and has become an optimal solution frequently implemented by biomaterials producers. An example of a successful hybrid modification is the surface coating of acetylated or thermally treated wood. The synergic effect of the reduced shrinkage/swelling of the bulk substrate and water protecting coating substantially reduces stresses of the coating film, thus preventing it from cracking. As a consequence, a façade surface remains intact for a longer period of time by preserving its original attractive appearance. Other examples of hybrid modifications implemented at an industrial scale are presented in Table 2.3.

Table 2.3 Hybrid modifications of bio-based materials suitable for the use in building façades

Constitutive treatments	Thermal treatment + water borne penetrating oil impregnation	Acetylation + surface coating	Oil impregnation + biofilm	Melamine impregnation + thermal treatment
Affected properties	• Improved dimensional stability • Better durability • Lower equilibrium moisture content	• Improved durability • High-dimensional stability • Change of the surface colour • Additional protection of the surface	• Increased resistance to biotic and abiotic degradation • Self-healing	• Enhanced dimensional stability • Fire resistance • Improved resistance to weathering
Appearance after modification				

(continued)

Table 2.3 (continued)

Constitutive treatments	Thermal treatment + water borne penetrating oil impregnation	Acetylation + surface coating	Oil impregnation + biofilm	Melamine impregnation + thermal treatment
Appearance after 1 year of natural weathering				

Benefits obtained by merging different materials and treatments are highly useful in addressing design limitations and biomaterial deficiencies. If properly implemented, hybrid modifications can help reduce the environmental burden and economic cost of façades. It has to be stated, however, that some modification processes cannot be merged or may induce undesired changes of other material properties. An example may be the increase of the brittleness of a biomaterial after certain hybrid modifications that affect its machinability or paintability. For this reason, special attention should be directed towards selecting appropriate treatment combinations and extensive quality control of the hybrid modification processes.

2.3.6 Bio-Based Composites

"Composite" is a term used to categorize materials merged with other materials possessing different structures or compositions. According to Rowell (2005), the key advantages of bio-based composites are:

- possibility to utilize waste from wood processing
- utilizing smaller trees
- removing material defects and deficiencies
- creating more uniform materials that are usually stronger than solid wood
- freedom in the shaping and design.

Bio-based Panels

The most widespread wood composites include glue-laminated beams, cross-laminated timber (CLT), plywood, oriented strand boards (OSB), particleboards, and fibreboards, among the others (Curling and Kers 2017). Not all of these are suitable to be used as façade elements. Tricoya® panel products, made from Tricoya® wood elements, are a groundbreaking construction material. In panel form, Tricoya® is opening new markets where wood-based panels would never have been considered before, such as wet interiors, kitchen carcasses, art installations window components, door skins, and building façades (Fig. 2.11). Tricoya® panels demonstrate significantly enhanced durability and exceptional dimensional stability. Tricoya® wood chips exhibit the same sustainable qualities such as longer lifespan and CO_2 sequestration, as its sister product Accoya® solid wood. Tricoya® is also guaranteed for 50 years above ground and 25 years in ground or fresh water due to its outstanding performance and properties.

Composites produced from alterative ligno-cellulosic materials, such as flax, hemp, straw, reed, wool, grass, bamboo, or rattan, are currently used in building interiors (e.g., flooring and siding) or as insulation in walls and roofs. Nevertheless, bamboo claddings (Fig. 2.12) and straw roofs (Fig. 2.13) have been recently frequently used in the building sector due to their unique and attractive appearance, sustainability, cost-effectiveness as well as the local identity (Knapic et al. 2017; Kotradyová 2015).

Fig. 2.11 Use of bio-based panels (Accoya® and MEDITE® TRICOYA® EXTREME) in the building façade

Fig. 2.12 Use of bamboo as a building façade material. Photograph courtesy of Kul-bamboo

Wood–Cement Composites

Inorganic materials proved to be a highly valuable component to be combined with natural materials. Gypsum–wood boards or cement–wood are examples of composites successfully utilized in the construction sector. These possess a high-dimensional stability, high durability against biotic and abiotic factors as well as high resistance against fire (Jorge et al. 2004). The addition of biomaterials reduces composite density and therefore makes the construction lighter. To produce panels, it is possible to use wood residuals, including waste from demolitions or

Fig. 2.13 Use of straw as the roof cover in a modern building

wood preserved even after its service. Similarly, recycled fly ash (residual from the combustion) can substitute up to 30% of the cement. Biomaterials other than wood are also used to produce composites, for example, palm, rattan, or bamboo, among the others. Cement–wood panels are frequently utilized as a siding of building façades. Even when compared to other bio-based solutions, cement-based façades are more expensive, heavier, and more difficult to assemble.

Wood–Plastic Composites

Wood–plastic composites (WPCs) are complex materials manufactured from different resins and wood powder/flour used as a filler (Kers and Ormondroyd 2017). Thermoset and naturally derived resins, such as polyethylene (PE), polypropylene (PP), polyvinyl chloride (PVC), and polyvinyl alcohol (PLA), are used in the majority of currently offered WPC products on the market. Wood–plastic composites have gained great interest of the resource-intensive building industry; however, according to Friedrich and Luible (2016), reliable technical data regarding application-oriented properties are still missing. On the contrary, WPCs are perceived by customers as maintenance free and have an excellent reputation regarding their durability and environmental friendliness (Morrell and Stark 2006). Recent studies performed by Turku et al. (2018) revealed that WPC weathering performance (e.g., changes in tensile strength and flexural properties) is influenced by its chemical composition. New improvements in WPC production are related to optimization of the manufacturing processes (extrusion, injection, or compression moulding) and the WPC composition (including the use of modified wood, non-wood fibres, nanoparticles, or fire retardants) (Gardner et al. 2015).

An important concern regarding the WPC is its environmental impact that may vary depending on the composite configuration and end-of-life scenario. In general, use of petroleum-based polymers results in a highly negative environmental impact. Conversely, renewable resource-based and biodegradable polymers with a high share of the wood filler are more environmentally friendly. Further reduction of the environmental impact can be achieved when wood used in WPCs comes from primary production side streams or is recovered from wood products (Schwarzkopf and Burnard 2016).

High-Pressure Laminate

High-pressure laminate (HPL) is a flat panel consolidated under heat and high pressure. It is made of wood-based layers impregnated with resin in a wide range of colours, finishes, and patterns. High-pressure laminate panels contain up to 70% natural fibres and do not require frequent maintenance after installation. Relatively high weather protection is provided by a coating with acrylic or polyurethane resins. An example of a building façades covered by the HPL is presented in Figs. 2.14 and 2.15.

Fig. 2.14 Use of HPL as a building façade. Photograph courtesy of Ewa Osiewicz

Fig. 2.15 Use of HPL in a building façade of Basket Apartments in Paris (OFIS Architects)

Engineered Wood–Glass Combination

The latest trend in building façade design is to provide multi-functionality and high energy efficiency at the same time. Particularly, the use of renewable materials with low environmental impact and attractive natural appearance, such as wood, coupled with large glazed areas, has recently gained increased interest (Tapparo 2017). An example of such a façade system, where timber load-bearing elements are merged with a protecting glass, is presented in Fig. 2.16. In this case, the glass layer protects the biomaterial from the direct wetting by the rain brought by wind as well as filters UV radiation present in the sunlight. As a result, the surface weathering kinetics are minimized, and the original biomaterial appearance is preserved. This is in line with the biophilic design approach, while assuring desired wood appearance not altered by environmental factors. Engineered wood–glass combination (EWGC) is thus a highly interesting solution in the modern architecture, especially where active and adaptive envelopes are desired.

2.3.7 Green Walls and Façades

Implementing greenery as the integral part of a building façade is a great solution to positively impact human well-being and to increase satisfaction of city occupants. Diverse configurations and implementations of this paradigm have been proposed by architects, creating a new trend of "living walls" or "vertical gardens". In general, such installations can be divided in two categories as described below.

Green Wall

Green wall is a part of the building intentionally covered by the living vegetation where plants are distributed on the whole surface of the façade. In this case, both the grooving media (usually soil) and the dedicated irrigation system are spread and

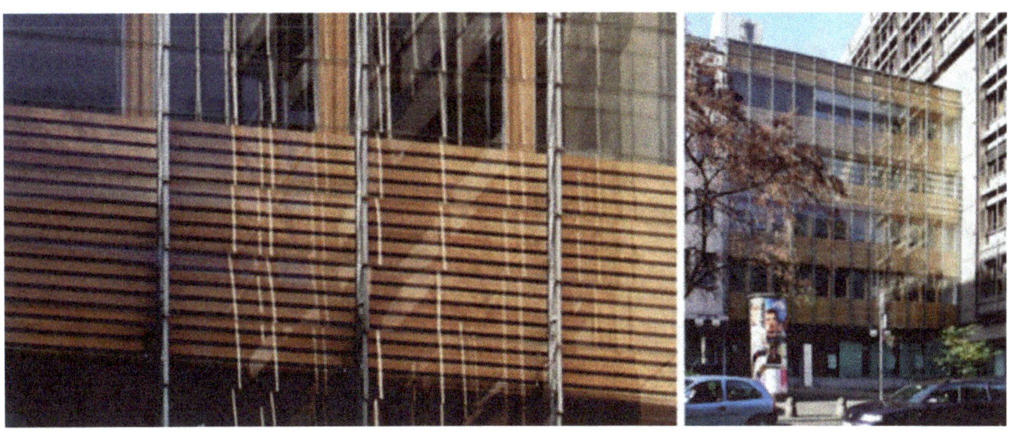

Fig. 2.16 Combination of the glass and biomaterials as a composite building façade of Bayerische Vereinsbank in Stuttgart (arch. Behnisch and Sabatke)

Fig. 2.17 Green wall implemented as a building façade

cover the whole area of the green wall (Fig. 2.17). Living (green) walls require specific supporting elements, growing substrate and efficient watering system. An important positive effect of the green wall is its capability to maintain consistent temperature and relative humidity on the inside of the building. Moreover, such installations decrease the wall temperature during summer time and provide thermal insulation in winter. Living plants offer shade, improve air quality, and dampen the effects of wind and noise. In some cases, greenery may cover a top of the building, thus creating so-called green roofs, as presented in Fig. 2.18.

Green Façade

In contrast to the green wall, the soil container necessary for plants to grow on the green façade is located at the base of the building façade. The plants covering the wall are therefore climbing on its face creating an external layer of vegetation.

Fig. 2.18 Green roof

Fig. 2.19 Green façade of Nagoya City Science Museum (arch. Nikken Sekkei)

The main challenge of green façades is their maintenance and investment costs as well as their installation (Besir and Cuce 2018). However, as highly attractive in terms of architectural and aesthetical aspects, green façades and walls are interesting alternatives for urban buildings of the future (Fig. 2.19).

2.4 Environmental Impact and Sustainability

2.4.1 Environmental Assessment

To provide solid evidence for supporting policy decisions, such as policies to encourage building with wood (particularly versus the use of non-renewable materials), the objective assessments of environmental impacts should be used. The claimed benefits of using renewable materials compared to non-renewable materials are backed by strong evidence when the whole life cycle of materials is considered. The life cycle of the renewable materials can reach the closed loop leading to closing the biological and technical metabolism (Fig. 2.20), while the life cycle of non-renewable materials cannot (Fig. 2.21). Benefits of renewable materials can be supported by an objective environmental impact assessment. The life cycle assessment (LCA) that considers the use and disposal as well as the reuse of materials and products is an objective measure of the environmental impacts of materials and products in their life cycle.

The LCA is a tool that has been developed to analyse and quantify the environmental burden associated with the production, use, and disposal of a product (Hill 2011). Furthermore, the LCA enables comparison of the environmental impacts of different products (Audenaert et al. 2012; Ding 2008; Forsberg and Von Malmborg 2004).

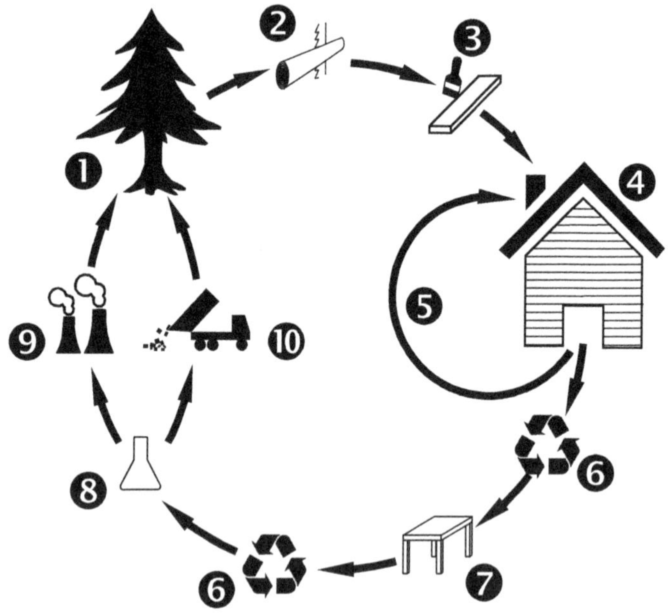

Fig. 2.20 Life cycle of renewable materials: ❶ harvesting, ❷ primary processing, ❸ secondary processing, ❹ use phase, ❺ reuse, ❻ recycling, ❼ second use phase, ❽ cascading to tertiary use, ❾ energy generation, ❿ landfilling, closing the biological and technical metabolism

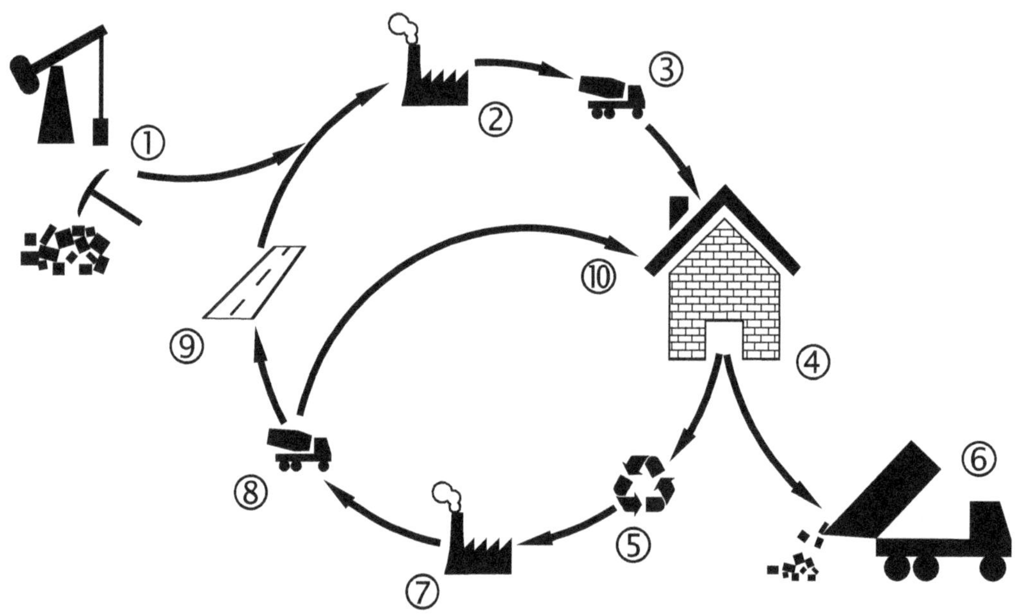

Fig. 2.21 Life cycle of non-renewable materials: ① extraction of raw material, ② manufacturing, ③ processing, ④ use phase, ⑤ recycling, ⑥ landfilling/waste production, ⑦ secondary manufacturing, ⑧ secondary processing for reuse, ⑨ second use phase, ⑩ second reuse phase

LCA Methodology

The LCA methodology is defined in ISO 14040 (2006a) and ISO 14044 (2006b). The most common methodologies to classify, characterize, and normalize environmental effects are focused on the following environmental impact indicators:

- acidification
- eutrophication
- thinning of the ozone layer
- various types of ecotoxicity
- air contaminants
- resource usage
- greenhouse gas emissions.

The LCA analysis is conducted by defining the goal and scope of the analysis that include the system boundaries and the functional unit. When materials are compared only until installed into a building, the system boundary is defined as "cradle to gate". Here, the environmental impacts are evaluated from the point of manufacture of a specific product in a factory to the point at which it leaves the facility. This corresponds to modules A1-A3 in the European Standard EN 15804 (2012). It provides the most accurate LCA because this phase of a product life cycle involves the fewest assumptions and the data gathering process is relatively straightforward. However, a low-impact product, as determined through a cradle-to-gate analysis, may require a lot of maintenance during the in-service phase of the life cycle, or there may be serious environmental impacts associated with its disposal. A full appreciation and understanding of the environmental impacts associated with a product choice therefore require the entire life cycle to be considered (Fig. 2.22). This invariably introduces a higher level of uncertainty into the process because there may be aspects of the life cycle that are not well understood, thus requiring assumptions to be made. These assumptions may have a very significant impact on a LCA, and a bias may be introduced if different products are being compared.

Furthermore, recycling and disposal may be analysed as well. The purpose of the LCA may be simply to report on the environmental burdens associated with a product or process, referred to as an attributional LCA, or to examine the consequences of changing various parameters or adopting different scenarios, referred to as a sequential LCA (Frischknecht and Stucki 2010; Gala and Raugei 2015).

The initial step in the LCA is also a determination of the subject of the LCA, that is, a declared unit or a functional unit. When cradle-to-gate is the system boundary of the analysis, it is referred to as the declared unit. When the analysis additionally includes other parts of the life cycle, it is referred to as the functional unit. In addition, the timescale included in the study and the allocation procedures are defined in the first step of the LCA.

When the goal and scope are defined, the life cycle inventory (LCI) phase of the analysis is performed. It requires a compilation of all information about the selected process. All material and energy inputs and outputs are quantified. This process is

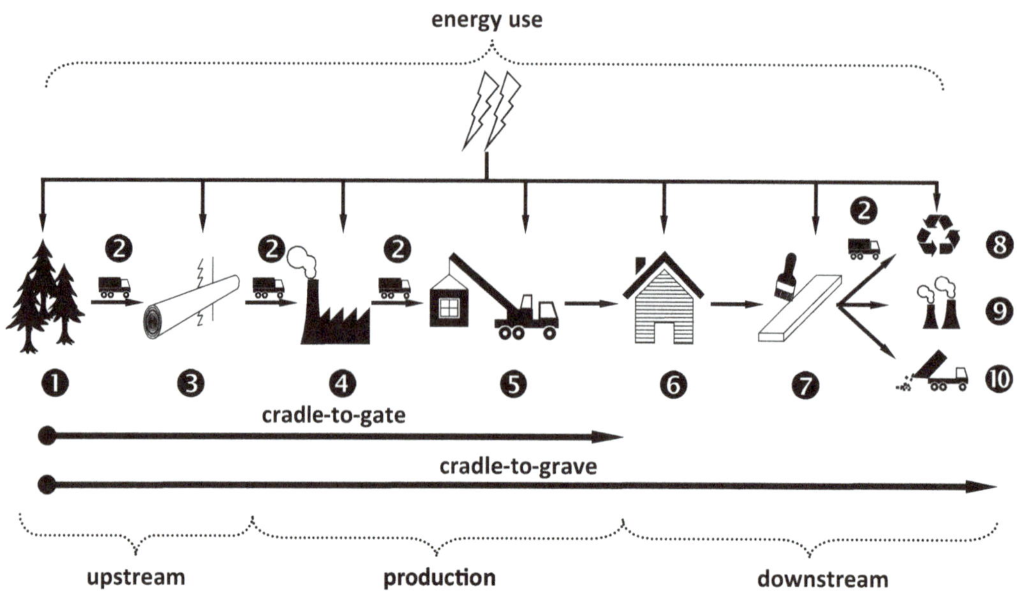

Fig. 2.22 Cradle-to-gate and cradle-to-grave concepts as the LCA system boundaries: ❶ harvest of raw material, ❷ transport, ❸ primary processing, ❹ secondary processing, ❺ construction/assembling, ❻ use phase, ❼ maintenance, ❽ recycling/reuse, ❾ energy generation, ❿ landfilling

divided into different life cycle stages, including manufacture, service life, end-of-life, and disposal. Data fall into two principal categories: primary (foreground) and secondary (background) data.

The LCI phase is followed by the life cycle impact assessment (LCIA) phase, and the environmental burden is quantified. The impact categories selected should provide useful information about the product or process while considering the goal and scope of the study. When selecting the impact categories, it is also necessary to select characterization factors, which are the units used to report each environmental burden.

LCA in the Wood Sector

The reported LCA studies of primary wood products mostly dealt with cradle-to-gate approach. This is due to the lack of data related to use phase maintenance, repair, refurbishment/replacement, as well as to deconstruction, demolition, waste processing, reuse, recovery, and recycling.

Kutnar and Hill (2014) discussed the environmental impacts of primary wood products and included a review of the LCA studies in the wood sector. They concluded that the research of timber processing and the resultant products focuses more on the interactive assessment of process parameters, developed product properties, and environmental impact, including recycling and disposal options at the end of the service life.

The fossil fuel consumption, potential contributions to the greenhouse effect, and quantities of solid waste tend to be minor in wood products compared to competing products that are used in the building sector (Werner and Richter 2007). However, impregnated wood products tend to be more critical than comparable products with respect to toxicological effects and/or photogenerated smog depending on the type of preservative. Bolin and Smith (2011a, b) compared environmental impacts related to borate-treated lumber and alkaline copper quaternary (ACQ)-treated lumber used for decking with a cradle-to-grave life cycle assessment. When compared to galvanized steel framing, the impacts of borate-treated lumber framing were approximately four times lower for fossil fuel use, 1.8 times lower for GHGs, 83 times lower for water use, 3.5 times lower for acidification, 2.5 times lower for ecological impact, 2.8 times lower for smog formation, and 3.3 times lower for eutrophication. The cradle-to-grave life cycle assessment of ACQ-treated lumber used for decking and façades was performed with the assumption that the ACQ decking has a service life of 10 years and that it is demolished and disposed in a solid waste landfill after the end of use. The study included the comparison with the wood–plastic composite (WPC) decking, which is the main alternative product to the ACQ decking. For the WPC, it was also assumed that the service life is 10 years. In both compared decking materials, maintenance, such as chemical cleaning and refurbishing, was not included in the LCI. The results of the cradle-to-grave life cycle assessment showed that ACQ-treated lumber impacts were fourteen times lower for fossil fuel use, almost three times lower for GHG emissions, potential smog emissions, and water use, four times lower for acidification, and almost twice lower for ecological toxicity when compared with WPC decking. Impacts were approximately equal for eutrophication.

The preservation or wood modification is extending the service life of materials. Hill and Norton (2014) compared different wood modification treatments. They defined the carbon neutrality—the point at which the benefits of life extension compensate for the increased environmental impact associated with the modification. Increased maintenance intervals of modified wood products help to lower the environmental impacts of the modified wood in the use phase.

2.4.2 Measures of Environmental Profiles

Environmental impacts of materials can be objectively compared when adequate guidelines are followed. ISO 14025 (2006) describes the procedures required in order to acquire Type III environmental product declaration (EPD). This allows comparability of environmental performance between products. The EPD is based on the principle of developing product category rules (PCR), which specify how the information from a LCA is to be used to generate the EPD.

The EPDs developed in Europe are mostly based on the PCR for "wood materials", which was released by the German Institute for Construction and Environment (Institut Bauen und Umwelt e.V.) in November 2009. PCR outlines five impacts on the environment: global warming potential, acidification potential, eutrophication potential, smog potential (photochemical oxidation), and ozone depletion potential (ozone layer depletion). ISO 14025 demands reporting of the

Table 2.4 Environmental product declaration programmes in Europe (summarized from Suttie et al. 2017)

	IBU: created in Germany in 2006; includes 41 categories
	BRE EN 15804 EPD: created in UK in 1999; includes all construction products
	EPD Norge: created in Norway in 2002; includes 19 categories
	EPD Environdec: created in Sweden in 2007; includes 13 categories
	Inies—Fiche de déclaration environnementale et sanitaire (FDES) created in France in 2004

environmental impacts of the production phase (cradle to gate) of the life cycle. The standard allows for other life cycle stages, such as the in-service stage and the end-of-life stage, to be included (but they are not compulsory). There has been a range of EPD programmes (Table 2.4) initiated since the publication of ISO 14025 (Del Borghi 2013). At the same time, a large number of PCRs were published. These PCRs, however, are not completely in agreement with each other (Subramanian et al. 2012).

The environmental performance of products that are relevant in the construction sector is also a subject of other standards. ISO 21930: 2017 provides the principles, specifications, and requirements to develop a PCR and EPD for construction products and services, construction elements, and integrated technical systems used in any type of construction work. In Europe, however, the EN 15804 (2012) was introduced as an alternative. It defines a core PCR for building products in more detail than the preceding ISO 14025 (2006). Here, the life cycle stages are divided into modules. Modules A1-A3 cover the production stage, A4-A5 the construction process, B1-B7 the use stage, and C1-C4 the end-of-life stage. In addition, stage D is used to analyse the product "after-life". Suttie et al. (2017) describe the main environmental impacts associated with each of the modulus for bio-based building materials and discuss the carbon accounting and benefits of using timber as a substitute for construction materials with higher embodied energy.

Besides the EPDs that are defined in the ISO 14025 (2006) as Type III Environmental Declarations, there is also an environmental label Type I corresponding to ecolabelling. These labels are based on a multi-criteria approach indicating the overall environmental performance of a product. Type I environmental labels are

Table 2.5 Examples of Type I—ecolabels (Summarized from Suttie et al. 2017)

	Nordic Ecolabel (1989): set up by Nordic Council of Ministers; official ecolabel of the Nordic countries; More information: http://www.nordic-ecolabel.org/
	Blue Angel (2013): set up by German Federal Minister of the interior; More information: https://www.blauer-engel.de/en
	NF Environment (1991): set up by AFNOR certification; French ecolabel scheme; More information: http://www.ecolabels.fr/en/the-nf-environnement-mark-what-is-it

provided by several programs established and operated in line with the requirements of ISO 14024 (2018). The environmental criteria that are taken into account are, for example, energy usage, climate aspects, water usage, source of raw materials, use of chemicals, hazardous effluents, packaging, and waste, among the others. Some examples of ecolabels used in different European countries are provided in Table 2.5.

Building materials and their environmental impacts are important also for assessments at the building level. In this case, the analysis includes a comprehensive assessment of environmental impacts and in most cases encourages the use of EPDs. The three most popular building assessment certifications are presented in Table 2.6.

Table 2.6 Examples of building certification schemes

	BRE environmental assessment methodology (BREEAM)
	Leadership in energy and environmental design (LEED)
	DGNB or German sustainable building council

2.4.3 Circular Economy, Reuse and Recycling of Biomaterials

Circular Economy Concept

Climate change and awareness of needed actions to satisfy multiple aspects of sustainability have led to the development of several political strategies defined at the European Union level. The "Waste Framework Directive" published in 2008 was one of the first such actions taken. The objective of this strategy was to reduce waste generation as well as to encourage increased use of waste as a resource.

Secondly, the "Roadmap 2050" was published in 2011 aiming to provide a practical, independent, and objective analysis of pathways to achieve a low-carbon economy in Europe. The document described a strategy that is in line with the energy security, environmental, and economic goals of the European Union. By the year 2050, Roadmap 2050 aims to reduce greenhouse gas (GHG) emissions to at least 80% below the levels present in 1990.

The "Bioeconomy Strategy" that addresses the production of renewable biological resources and their conversion into vital products and bioenergy was published in 2012. It is structured around three pillars:

- investments in research, innovation, and skills
- reinforced policy interaction and stakeholder engagement
- enhancement of markets and competitiveness.

The forest-based sector is a key pillar of Europe's bioeconomy. Using wood products can contribute to significant CO_2 saving in terms of greenhouse gas emissions, embodied energy, and energy efficiency (Hill 2011). The three above-mentioned documents—Waste Directive, Roadmap 2050, and Bioeconomy Strategy—provide a prospect for the increased use of biomaterials in general, but especially in the construction sector. The increased use was further promoted by the Circular Economy Package published in 2018. The Circular Economy Package includes revised legislative proposals on waste to stimulate Europe's transition towards a circular economy which will boost global competitiveness, foster sustainable economic growth, and generate new jobs. The proposed actions should contribute to "closing the loop" of product life cycles through greater recycling and reuse and bring benefits for both the environment and the economy. The specific measures to promote reuse and stimulate industrial symbiosis are described with a special emphasis on turning a by-product of one industry into a raw material for another. Biomaterials are directly fulfilling aims of the Circular Economy, assuming that harmful chemicals are not involved along the life cycle.

The advantageous aspects of using biomaterials as building materials are also related to the list of goals as defined in the recently published "Research and Innovation Roadmap 2050—Sustainable and Competitive Future for European Raw Materials" (Reynolds 2018). The policy aims to secure a sustainable and competitive supply of raw materials, boost the sector's jobs and competitiveness, and contribute to addressing global challenges as well as the needs of the society. The priority areas and required activities are directed towards the supply of raw resources, production of raw materials, "closed loops", as well as innovative products and applications. A further opportunity for the development of the bio-material sector is finding solutions for substituting critical raw materials, along with the development of new bio-based products, such as composite materials.

Cascade Use of Resources

Reducing waste is a fundamental element in protecting the natural environment. The general concept in minimizing the amount of waste is based on "reduce–reuse–recycle" paradigm. The "reuse" is a preferred option and includes the transformation and the development of new products with minimal cost. Since the materials are used in their original form, efforts related to these conversions are minimized. An example of the material reuse is the manufacture of flooring or furniture from building cladding or the use of parts of demolished buildings in other structures. Another example is the Circular Pavilion in Paris designed by studio Encore Heureux, where the façade of the pavilion is crafted from 180 recycled wooden doors (Fig. 2.23). Even though this was produced as a work of art, it brought high public attention to the benefits of a low-waste, circular economy.

Reused engineered solid wood products, such as cross-laminated timber or glue-lam recovered from large structures, are a highly valuable source of construction materials. Such resources are of high graded quality, with the optimal hygroscopic properties and relaxed internal stresses. Recycled constructional wood can thus be used in the timber housing industry. Casa MAI—Modulo Abitativo IVALSA, presented in the Fig. 2.24, is a perfect example of the pioneer implementation of the reuse paradigm. It is an experimental transportable house constructed entirely from cross-laminated timber panels recovered from the building structure of the SOFIE project. MAI prototype was designed by DUOPUU and

Fig. 2.23 Recycled wooden doors used as a building façade

Wood building design lab of CNR-IVALSA within a research project on sustainable buildings carried out in collaboration with Ceii Trentino and with the support of Provincia Autonoma di Trento. The goal of the project was to design and build a small wooden prototype house (about 35 m^2 floor area) that can be built and prefabricated in a controlled area (factory), easily transported by truck to the final destination and finally assembled to form a real house in a few hours. The MAI prototype, when leaving the factory, was ready to be assembled and used: The heating–ventilation–air-conditioning system, electric installation, plumbing, interior finishing, lights, appliances, and furniture were already installed. All the construction phases in the production area have been optimized to reduce the waste of materials, control air and water pollution, and achieve a highly efficient building process with respect to the environment and the workers (Briani et al. 2012).

Recycling aims to convert waste timber into usable products. There are three distinct types of recycling—direct, indirect, and energy recovery. Direct recycling includes all processes where one sort of timber products is recycled to other sorts of timber products, usually to laminated timber, wood-based panels, or wood–plastic composites. Indirect recycling changes timber products into other types of products, such as animal bedding, landscape mulch, cement boards, or compost. Energy recovery is a process aiming to make use of embedded calorific energy stored in waste biomaterials. Processes of energy recovery are relatively simple and include preparation of fuel (chips, pellets, or briquets), combustion, co-firing, co-generation, pyrolysis, or gasification.

Fig. 2.24 Casa MAI as an example of a building constructed with reused cross-laminated panels. Image courtesy of Romano Magrone and Paolo Simeone—DUOPUU

A sequential use of a certain resource for different purposes is termed "cascade use". It enables using the same material unit in multiple high-grade material applications occurring consecutively from the most complex to the simplest. The ultimate stage of the cascade use is usually energy conversion. The majority of wood-based products at the end of service life are in the state that allows straightforward further use (Fig. 2.25). Study of Höglmeier et al. (2013) demonstrated, for example, a great potential for cascading of wood recovered from building deconstruction. By utilizing recovered wood, the time span of carbon storage in the wood products increases and consequently delays the contribution to the greenhouse effect. Unfortunately, in many real-world cases the recovered wood is currently considered as not usable for cascade use and is simply burned or landfilled.

The amount of CO_2 wood releases into the atmosphere during burning or in-field decomposition is comparable to the amount absorbed by the tree during its growth. However, even if incineration of wood products at the end of life provides various environmental benefits, the research demonstrated that the use of forest residues in manufacturing particleboards is more sustainable than when they are used as fuel (Rivela 2006). It was shown that manufacturing particleboards from waste wood produces per 1 ton of the final product 428 kg CO_2 equivalent less than particleboard manufactured from fresh wood. Cascading through several life cycles prior to incineration is therefore a fairly better option for the end-of-life of bio-based materials. Table 2.7 summarizes the suitability of diverse waste processing

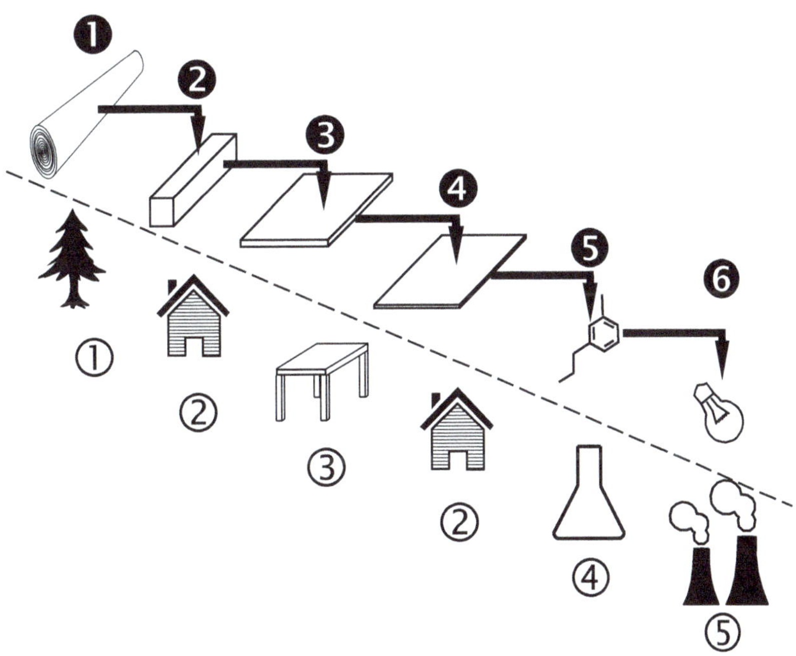

Fig. 2.25 Cascade use of biomaterials: ❶ log, ❷ large dimensions solid or engineered timber assortments, ❸ strand-/particle-based composites, ❹ fibre-based composites, ❺ chemicals, ❻ energy, ① resource extraction, ② first life cycle, ③ second life cycle, ④ chemicals processing, ⑤ energy generation

Table 2.7 Processing technologies for bio-based building residuals

Processing technology	Feedstock flexibility	Conversion efficiency	Market value of product
Combustion	High	Low	Low
Digestion	Low	Medium	Medium
Fermentation	Low	Medium	High
Pyrolysis	High	Medium	Medium
Gasification	Medium	Medium	Medium
Platform molecules	Medium	Medium	High
Liquefaction	Medium	Low	High
Composites manufacturing	High	High	High
Animal bedding	High	Medium	Low
Pelletizing	High	High	High
Insects conversion	Medium	Medium	High
Fungal conversion	Medium	Medium	High

technologies that are available for bio-based building materials. The restrictions identified are related to the feedstock flexibility, efficiency of the process, and the value of final products on the market. It should be stated that some of the listed technologies are still at the development stage; however, their validation and upscaling is only a matter of time. It is expected that intelligent concepts for the reuse and recycling of valuable materials at the end-of-service life will reduce the amount of landfilled waste. So far, landfilling is a most frequent path of the waste transformation, even if it is recognized as a less than optimal solution.

References

Attias N, Danai O, Ezov N et al (2017) Developing novel applications of mycelium based biocomposite materials for design and architecture. In: Final COST FP1303 conference "Building with bio-based materials: best practice and performance specification", Zagreb

Audenaert A, De Cleyn SH, Buyle M (2012) LCA of low-energy flats using the eco-indicator 99 method: impact of insulation materials. En Build 47:68–73

Besir AB, Cuce E (2018) Green roofs and façades: a comprehensive review. Renew Sustain En Rev 82(1):915–939

Bolin CA, Smith ST (2011a) Life cycle assessment of borate-treated lumber with comparison to galvanized steel framing. J Clean Prod 19:630–639

Bolin CA, Smith ST (2011b) Life cycle assessment of ACQ-treated lumber with comparison to wood plastic composite decking. J Clean Prod 19:620–629

Briani A, Simeone P, Ceccotti A (2012) MAI Ivalsa modular house. In: Proceedings of the 12th world conference on timber engineering, Auckland, New Zealand

Brischke C (2017) Moisture performance. In: Jones D, Brischke C (eds) Performance of bio-based building materials. Woodhead Publishing, pp 277–285

Buchanan A, Östman B, Frangi A (2014) White paper on fire resistance of timber structures. NIST GCR, 15-985. https://doi.org/10.6028/NIST.GCR.15-985

Burnard MD, Nyrud AQ, Bysheim K et al (2017) Building material naturalness: perceptions from Finland, Norway and Slovenia. Indoor Built Env 26(1):92–107

CEN/TS 15679 (2007) Thermal modified timber—definitions and characteristics

Curling S, Kers J (2017) Panels. In: Jones D, Brischke C (eds) Performance of bio-based building materials. Woodhead Publishing, pp 52–61

Del Borghi A (2013) LCA and communication: environmental product declaration. Int J Life Cycle Assess 18(2):293–295

Ding GK (2008) Sustainable construction—the role of environmental assessment tools. J Env Manage 86(3):451–464

ECE/TIM/SP/33 (2013) Forest products annual market review 2012–2013. UNECE/FAO Forestry and Timber Section, Geneva. ISBN 978-92-1-117070-2

EN 350 (2016) Durability of wood and wood-based products. In: Testing and classification of the durability to biological agents of wood and wood-based materials. European Committee for Standardization, Brussels, Belgium

EN 15804 (2012) Sustainability of construction works—environmental product declarations—core rules for the product category of construction products. European Committee for Standardization, Brussels, Belgium

Esteves B, Pereira H (2009) Wood modification by heat-treatment: a review. BioResources 4 (1):370–404

European Commission (2008) Directive 2008/98/EC on waste

European Commission (2011) A roadmap for moving to a competitive low carbon economy in 2050

European Commission (2012) Innovating for sustainable growth—a bioeconomy for Europe

European Commission (2018) Circular economy package

Foliente GC (2000) Developments in performance-based building codes and standards. Forest Prod J 50(7):12–21

Fortini A (2017) The latest design trend: black and burned wood. The New York Times Style Magazine, 24.09.2017, p 176

Forsberg A, Von Malmborg F (2004) Tools for environmental assessment of the built environment. Build Env 39(2):223–228

Friedrich D, Luible A (2016) Standard-compliant development of a design value for wood–plastic composite cladding: an application-oriented perspective. Case Stud Struct Eng 5:13–17

Frischknecht R, Stucki M (2010) Scope-dependent modelling of electricity supply in life cycle assessments. Int J Life Cycle Assess 15(8):806–816

Gala AB, Raugei M (2015) Introducing a new method for calculating the environmental credits of end-of-life material recovery in attributional LCA. Int J Life Cycle Assess 20(5):645–654

Gardner DJ, Han Y, Wang L (2015) Wood-plastic composite technology. Curr For Rep 1:139–150

Gerard R, Barber D, Wolski A (2013) Fire safety challenges of tall wood buildings. Arup North America Ltd. San Francisco, CA, and Fire Protection Research Foundation Quincy, MA, USA

Gérardin P (2016) New alternatives for wood preservation based on thermal and chemical modification of wood—a review. Ann For Sci 73(3):559–570

Greef JM, Brischke C (2017) Reed. In: Jones D, Brischke C (eds) Performance of bio-based building materials. Woodhead Publishing, pp 124–128

Hill C (2006) Wood modification chemical, thermal and other processes. Wiley, Chichester, England

Hill C (2011) An introduction to sustainable resource use. Taylor and Francis, London

Hill C, Norton A (2014) The environmental impacts associated with wood modification balanced by the benefits of life extension. In: European wood conference on wood modification, Lisbon, 8 p

Hodsman L, Smallwood M, Williams D (2005) The promotion on non-food crops. The National Non-Food Crops Center, IP/B/AGRI/ST/2005–02

Höglmeier K, Weber-Blaschke G, Richter K (2013) Potentials for cascading of recovered wood from building deconstruction-a case study for south-east Germany. Resour Conserv Recycl 78:81–91

http://www.servowood.eu. Accessed 05 Oct 2018

ISO 14040 (2006) Environmental management—life cycle assessment—principles and framework. International Organisation for Standardization, Geneva, Switzerland

ISO 14044 (2006) Environmental management—management—life cycle assessment—requirements and guidelines. International Organisation for Standardization, Geneva, Switzerland

ISO 14025 (2006) Environmental labels and declarations—type III environmental declarations—principles and procedures. International Organisation for Standardization, Geneva, Switzerland

ISO 21930 (2017) Sustainability in buildings and civil engineering works—core rules for environmental product declarations of construction products and services. International Standards Organization, Geneva, Switzerland

ISO 14024 (2018) Environmental labels and declarations—type I environmental labelling—principles and procedures. International Organisation for Standardization, Geneva, Switzerland

Jones N, Mundy J (2014) Environmental impact of biomaterials and biomass. BRE Press. ISBN: 978-1-84806-371-6

Jorge FC, Pereira C, Ferreira JMF (2004) Wood-cement composite: a review. Holz Roh Werkst 62 (5):370–377

Kellert S, Heerwagen J, Mador M (2008) Biophilic design: the theory, science, and practice of bringing buildings to life. Willey, Chichester

Kers J, Ormondroyd GA (2017) Wood plastic composites. In: Jones D, Brischke C (eds) Performance of bio-based building materials. Woodhead Publishing, pp 61–68

Knapic S, Bajraktari A, Nunes L (2017) Bamboo and rattan. In: Jones D, Brischke C (eds) Performance of bio-based building materials. Woodhead Publishing, pp 120–124

Kotradyová V (2013) Authenticity vs. artificiality: the interactions of human beingswith objects in the built environment—body conscious perspective by spatial design (Chapter X). In: Tzvetanova Yung S, Piebalga A (eds) Design directions: the relationship between humans and technology. Cambridge Scholars Publishing, pp 115–148

Kotradyová V, Kaliňáková B (2014) Wood as material suitable for health care and therapeutic facilities. Adv Mater Res 1041:362–366

Kotradyová V (2015) Material surface features in body conscious spatial design. Int J Contemp Arch 2(2):38–44

Kutnar A, Hill C (2014) Assessment of carbon footprinting in the wood industry. In: Muthu SS (ed). Assessment of carbon footprint in different industrial sectors, vol 2, (EcoProduction). Singapore, Springer, pp 135–172

Mansour E, Ormondroyd GA (2017) Wool. In: Jones D, Brischke C (eds) Performance of bio-based building materials. Woodhead Publishing, pp 128–141

Mazela B, Popescu CM (2017) Solid wood. In: Jones D, Brischke C. (eds) Performance of bio-based building materials. Woodhead Publishing, pp 22–39

Morrell JJ, Stark NM (2006) Durability of wood-plastic composites. Wood Des Focus 16(3):7–10

Navi P, Sandberg D (2012) Thermo-hydro-mechanical wood processing. CRC Press, Boca Raton

Odeen K (1985) Fire resistance of wood structures. Fire Technol 21(1):34–40

Östman B (ed) (2010) Fire safety in timber buildings—technical guideline for Europe. SP Report 2010:19. Stockholm, Sweden

Rautkari L, Kutnar A, Hughes M et al (2010) Wood surface densification using different methods. In: 11th world conference on timber engineering, Riva del Garda, pp 647–648

Réh R, Barbu MC (2017a) Flax. In: Jones D, Brischke C (eds) Performance of bio-based building materials. Woodhead Publishing, pp 98–105

Réh R, Barbu MC (2017b) Hemp. In: Jones D, Brischke C (eds) Performance of bio-based building materials. Woodhead Publishing, pp 105–112

Reinprecht L (2016) Wood deterioration, protection and maintenance. Wiley, Oxford, UK

Reynolds T (2018) Research and innovation roadmap 2050—sustainable and competitive future for European raw materials

Rivela B, Hospido A, Moreira T et al (2006) Life cycle inventory of particleboard: a case study in the wood sector. Int J Life Cycle Assess 11(2):106–113

Rowell (2005) Handbook of wood chemistry and wood composites. CRC Press

Rowell RM (2006) Acetylation of wood—journey form analytical technique to commercial reality. Forest Product J 56(9):4–12

Russell LJ, Marney DCO, Humphrey DG et al (2007) Combining fire retardant and preservative systems for timber products in exposed applications—state of the art review. Project no: PN04.2007, Forest and wood products R & D Corporation, Victoria

Schwarzkopf MJ, Burnard MD (2016) Wood-plastic composites—performance and environmental impacts. In: Kutnar A, Muthu SS (eds) Environmental impacts of traditional and innovative forest-based bioproducts, environmental footprints and eco-design of products and processes. Springer, Singapore

Subramanian V, Ingwersen W, Hensler C et al (2012) Comparing product category rules from different programs: learned outcomes towards global alignment. Int J Life Cycle Assess, pp 892–903

Suttie E, Hill C, Sandin G et al (2017) Environmental assessment of bio-based building materials. In: Jones D, Brischke C (eds) Performance of bio-based building materials. Woodhead Publishing, pp 547–591

Tapparo A (2017) Engineered wood glass combination—innovative glazing façade system. Master Thesis, KTH Royal Institute of Technology. ISBN 978-91-7729-476-4

Tellnes LGF, Ganne-Chedeville C, Dias A et al (2017) Comparative assessment for biogenic carbon accounting methods in carbon footprint of products: a review study for construction materials based on forest products. iForest 10:815–823

Teppand T (2017) Grass. In: Jones D, Brischke C (eds) Performance of bio-based building materials. Woodhead Publishing pp 150–156

Turku I, Kärki T, Puurtinen A (2018) Durability of wood plastic composites manufactured from recycled plastic. Heliyon 4(3):e00559

Walker P, Thomson A, Maskell D (2017) Straw. In: Jones D, Brischke C (eds) Performance of bio-based building materials. Woodhead Publishing, pp 112–119

Werner F, Richter K (2007) Wood building products in comparative LCA. A literature review. Int J LCA 12(7):470–479

Willems W (2014) The hydrostatic pressure and temperature dependence of wood moisture sorption isotherms. Wood Sci Technol 48:483–498

Chapter 3
Designing Building Skins with Biomaterials

We shape our buildings; thereafter they shape us.
Winston Churchill

Abstract This chapter presents several successful examples of biomaterial façade design. It discusses façade function from aesthetical, functional, and safety perspectives. Special focus is directed on novel concepts for adaptation and special functionalities of façades. Analysis of the structure morphologies and aesthetic impressions related to the bio-based building façades is supported with photographs collected by authors in various locations. Finally, particular adaptations and special functionalities of bio-based façades going beyond traditional building envelope concept are supported by selected case studies.

The world population is gradually increasing. In consequence, many new buildings will be erected in the near future to provide housing, services, and recreation. However, it is estimated that buildings are already responsible for 40% of the total energy consumption and 36% of the total CO_2 emissions (Herczeg et al. 2014). It is desired that the renovation and construction of buildings/infrastructure will be highly resource efficient by 2020. Recent trends in advanced material research have focused on the development of solutions optimized for specific applications that assure the expected properties and functionality over elongated service lives with minimized environmental impact and reduced risk of product failure.

Sustainability of bio-based materials is generally highly valued. There are two main groups of environmental benefits associated with the use of bio-based materials. The first group is associated with material production. Carbon dioxide is trapped in organic tissues of biomaterials and, as a result, does not contribute to the greenhouse effect and the climate warming. Processing of bio-based materials, although not always environmentally friendly (e.g., the use of chemical binders, high water use) represents, in general, the best circular economy practice. Namely, even if waste is generated (e.g., chips, sawdust), it becomes a raw material for the subsequent production cycles (e.g., particle board, OSB), as endorsed by the "waste to resources" principle. The second group of advantages is associated with the end-of-life of

A. Sandak et al., *Bio-based Building Skin*, Environmental Footprints and Eco-design of Products and Processes, https://doi.org/10.1007/978-981-13-3747-5_3

bio-based products. Biodegradability enables environmentally friendly decomposition of products and a return of chemical compounds back to the natural cycles.

3.1 Functions of Biomaterials in Buildings

In buildings, biomaterials perform various functions. Timber elements do not only constitute a load-bearing structure of buildings, but also form its skin that separates external and internal environments. Biomaterials also have an important aesthetic function that is much valued by the building users. While performing various functions, biomaterials must be safe and, wherever possible, adaptable to the needs of inhabitants.

Timber façade introduces many design challenges. Timber is renewable but also biodegradable, which leads to unique obstacles in design. Usually, façades are associated with durability and resistance when facing harsh weather conditions. However, biomaterials used in façades in some cases are less robust than non-bio-based materials. Still, if properly designed, façades constructed from biomaterials could be at least equally resistant as non-bio-based façades. Understanding all façade functions is essential for a successful application of biomaterials in building practices.

3.1.1 Façade as a Barrier and Interface Between the Outside and the Inside

From the energy conservation perspective, façades are barriers between interiors of buildings and the surrounding environment (also labelled as a filtering layer between the outside and the inside). In buildings, the internal environment (microclimate) has to be maintained within the certain limits to provide comfort for the occupants. In a temperate climate, the external environment is variable and changes substantially over the course of the year (four seasons). To conserve energy, it is crucial to minimize the energy transfer between the interior and the exterior of a façade while still maintaining a healthy and pleasant internal microclimate (Lovel 2013). Façades perform many functions in buildings. They provide a shelter for humans, adequate air exchange, light transmission, and a boundary between the public and the private (Herzog et al. 2004). Since they perform a variety of tasks, façades are considered by Boswell (2013) as technically one of the most complex features in architecture.

Functions of façades can be broadly divided into two groups. One group is associated with façade robustness and resilience against the external conditions, while the other relates to the proper maintenance of internal microclimate with minimized use of energy and resources. Building façade is expected to achieve

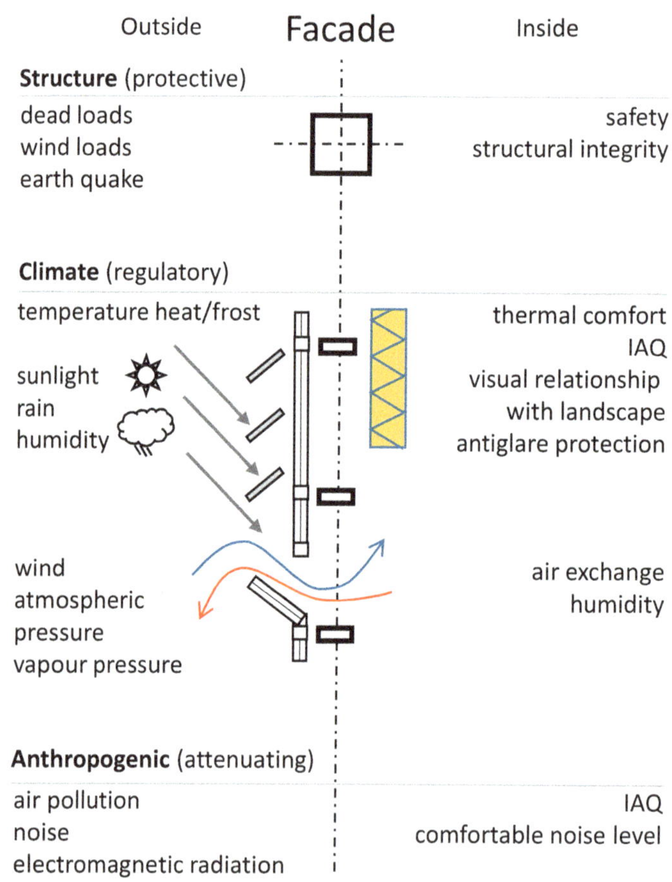

Fig. 3.1 Façade as a border between outside and inside of the building

many objectives: from a building's structural integrity (load bearing) to its proper relationship with the surrounding environment. In some occasions, façades must provide insulation (e.g., protect against heat escape, water ingress, noise), while in others they must ensure permeability (e.g., of air, daylight, view, or vapour). A brief review of factors influencing façade functions is summarized in Fig. 3.1 and presented below.

External Conditions

Façade designers cannot influence external conditions such as climate, pollution, and electromagnetic radiation. Such conditions vary widely depending on the location and orientation of a building. They are, nonetheless, important to consider. Some of them are constant, including orientation, ground-carrying capacity, and water level, while others are variable (diurnally or seasonally), for example temperature, solar radiation, atmospheric pressure, wind speed, and precipitation. It is challenging to design protective measures that are able to withstand such external conditions throughout the building lifetime. Still, impacts of climatic factors should be considered during the design phase of a building. The main factors to take into account include sun radiation and temperature change, rain and air humidity fluctuations, and influence of the wind.

Sun radiation includes infrared, visible, and ultraviolet light. Several parts of the solar spectrum, particularly UV, are contributing to the photodegradation of chemical constituents of materials, which changes the colour of their surface. Direct exposure of biomaterials to light increases their temperature. This is followed by the change in moisture levels that subsequently leads to dimensional distortions. On the other hand, if properly collected, sun radiation may provide ample amounts of energy to buildings. Since direct penetration of light may be irritating for building occupants and can rise the internal temperature, various sun shading systems are implemented (e.g., shutters, blinds, brise-soleil, lamellas).

Typically, normal variations in temperature do not significantly affect bio-based materials as they are accustomed to changing environmental conditions. However, if materials originate from climatic conditions that differ from the building location climate (e.g., when bamboo is used in a temperate climate), materials should be carefully studied to discover all potential incompatibilities (Kim et al. 2016). Water-absorbing elements (if present) can be prone to destruction from frost, as the frozen water—ice—expands. However, it is important to emphasize that the internal flexibility and cellular structure of bio-based materials make them more resistant for repetitive frost and thaw cycles compared to porous rigid materials, such as concrete or stone.

Rain penetration protection includes all measures taken to limit the water ingress into façades. When the ingress is permitted, water should be periodically allowed to evaporate completely. In case of bio-based cladding materials, this includes proper joint detailing that should be either open drained (as in ventilated façades) or overlapped, which is used to direct the water flow towards the façade base (joint transverse to the water flow) (Knaack et al. 2007). It is especially important to consider water ingress in the context of biomaterials, since many of them absorb and store large quantities of water. For that reason, an effective ventilation and air circulation seem to be some of the most important issues in this field. When bio-based materials are exposed to the intense precipitation, it is common to employ an external weatherproofing layer. The design of this layer enables easy renovation after a period of exposure without requiring changes of the underlying façade structure (Herzog et al. 2004). Closely related to precipitation levels are fluctuations in air humidity that can lead to large-dimensional distortions or even the destruction of façades. Bio-based materials absorb large quantities of water from the vapour in the air. Therefore, components in which an alternating moisture content is expected must enable predictable dimensional changes (Herzog et al. 2012). Occupied rooms should be protected from uncontrolled air exchange and draughts, especially if these are resulting from the high wind pressure on the façade. Proper design and subsequent craftsmanship of façades can protect buildings from strong winds. Ordinarily, joints in the external, exposed layer should be designed as overlapped or grooved to assure the proper seal. Airtightness is even more important in passive buildings, where the air exchange is provided by mechanical ventilation with the heat recuperation. In such cases, a leak in the building envelope may influence the building energy performance. The most common solution to this problem, especially in timber-framed buildings, is to wrap the building in a vapour permeable membrane (e.g., Tyvek by DuPont). This barrier must be positioned on the outside of the thermal insulation layers. A proper sealing of the membrane, junction around the corners, and the openings (windows and doors) become issues of major importance.

Internal Comfort Conditions (Including Indoor Air Quality)

Performance of bio-based materials can substantially contribute to the microclimate comfort by managing energy and mass (vapour) transfer. It is important to note, however, that user comfort also depends on the occupant preferences and characteristics, such as clothing, personality, or cultural background. For example, British citizens are comfortable with lower temperatures than citizens of other European countries (Stevenson and Baborska-Narozny 2018). Relevant factors, influencing occupant comfort, are listed below and briefly discussed in the context of bio-based material use.

Temperature

Internal air temperature is a basic factor influencing occupant comfort. It is important to be aware that it could be measured in different ways (i.e., taking the humidity in the account or not) which results in distinct outputs (e.g., dry-bulb temperature, wet-bulb temperature, mean radiant temperature). The most important façade feature related to user thermal comfort is the provision of proper insulation to prevent overheating or/and cooling. Thermal performance of bio-based materials is related to their internal cellular and fibrous structure. Timber acts as a thermal insulator while simultaneously providing a suitable internal surface temperature. Timber also protects from thermal bridges, as it is one of the very few available materials capable of both load bearing and insulation. Wood volumetric change due to heat is minimal; thus, timber is considered to be a good structural material in many circumstances, for example, in solid timber structures, glued arrangements (e.g., glulam or cross-laminated timber (CLT)), or framed solutions (Herzog et al. 2012). Timber, however, possesses a limited potential to act as a thermal building mass in comparison with brick or concrete, since it has a relatively low density and a high heat capacity. Other processed bio-based composites (e.g., wood-based foams or fibrous insulation materials) are more promising for the regulation of the building internal climate, since they are better insulators. Nowadays, insulation materials are manufactured from numerous types of bio-based resources, where some innovative solutions based on plant residuals offer up to 20% higher insulation levels than traditional materials (Sid 2018). A wide variety of innovative resources includes fungal mycelium that is preferably grown (cultivated) than produced (converted) (Fig. 3.2). The new types of bricks are innovative combination of stalk waste and living mushrooms. They are lightweight, low cost, and sustainable. Such material can be decomposed in 60 days and is an interesting alternative to wasteful linear economy. Similar experiments are conducted by several independent research groups aiming to improve growing process and

Fig. 3.2 Bricks made of fungal mycelium. Image courtesy of The Living

thermal performance of the mycelium materials (Attias et al. 2017; Xing et al. 2018). It is estimated that such locally cultivated and processed materials could contribute to the reduction of up to 50% of the building's total embodied energy across the whole life cycle (Sid 2018).

Relative Humidity (RH)

Relative humidity (RH), as well as vapour pressure and migration, is an important component of the indoor air quality. It is important to keep the level of relative humidity within certain limits (usually approximately 30–45% at 20 °C in the office environment) to provide healthy and comfortable living and working conditions. Recommended humidity levels may vary depending on the air temperature. Moisture in the air enables human gaseous exchange system to operate properly, supports the functioning of mucous membranes (either in the eyes' conjunctiva or in the throat), and enhances perception of comfort and well-being. In addition, appropriate level of relative humidity restrains the survival of various viruses (Noti et al. 2013).

Water vapour migrates from the environment with higher vapour pressure to the environment with lower vapour pressure. By being partly moisture permeable, porous façades (also called "breathable") are able to balance this internal/external difference. In general, bio-based materials possess good properties of vapour diffusion that assure microclimatic comfort, and also protect against vapour accumulation in rooms, thus blocking subsequent potentially dangerous condensation. Properties of vapour diffusion through the bio-based materials are inherent properties of the material structure itself. Bio-based materials are generally hygroscopic; that is, they retain water molecules until an equilibrium state of water content related to the RH of the ambient air is reached (ASHRAE 2004). This positively influences internal air quality, as it creates a buffer mechanism that balances rapid changes in relative humidity. Air can absorb water vapour until it reaches its saturation point, which depends on the temperature. The relative humidity has to be always considered in conjunction with room temperature.

Draughtiness

ASHRAE Standard 55-2004 defines draught as an "undesirable local cooling of the body due to air movement" (ASHRAE 2004). Draughts can result from improper installation of an air insulation layer, from mechanical ventilation forcing air into the occupied space at a too high speed, or from natural ventilation resulting from the pressure difference between the wind- and leeward sides of façades. The use of bio-based materials does not substantially affect occupant comfort related to draughts, unless the external bio-based layer installation is faulty or leaks occur.

Toxins and Odours

Many symptoms of the sick building syndrome, such as headache, nausea, drowsiness, dizziness, and nasal congestion may result from the use of materials emitting hazardous substances. In this context, bio-based materials might pose two groups of risks. The first is associated with chemicals used to protect (or aesthetically improve) solid wood products (e.g., protection from dirt or water ingress), including impregnates, coatings, paints, and stains. The second is associated with the use of adhesives in engineered wood products, such as CLT, particle boards, OSB, or plywood. The most frequently used adhesive in wood industry is the urea formaldehyde (UF) resin. Some formaldehyde molecules are not cross-linked with the glue bond, may be released into the room in the form of vapour, and can thus be absorbed by occupants through inhalation. A formaldehyde concentration level exceeding 0.1 ppm may cause coughing, wheezing, vomiting, and skin irritation (Raja and Sultana 2012). More importantly, however, formaldehyde inhalation can cause cancer, according to the International Chemical Safety Card ICSC: 0275 (ICSC 2012). Other commonly used adhesives include melamine formaldehyde (MF), phenol formaldehyde (PF), and methylene diphenyl diisocyanate (MDI). The trend for using bio-based binders led to incorporating cellulose, proteins, lignin, tannins, and fatty oils into various innovative adhesives. The biggest challenges that bio-based adhesives must overcome to compete with synthetic adhesives are related not only to their performance but also to their economic and production requirements (Frihart 2016). Nevertheless, the recent review published by Ikei et al. (2017) shows several positive effects of olfactory stimulation of humans by wood-derived substances.

Amount and Quality of Light (Lighting Environment) and Redirection of Daylight.

Suitable levels of room daylight facilitate occupant comfort. Natural illumination levels depend on numerous factors including the position of the sun, the orientation of the building, the size of the openings, the depth of the reveal, and the colour of internal surfaces. Daylight distribution and redirection is an issue concerning all rooms but especially those that are illuminated on a single side (Herzog et al. 2004). The level of daylight in the room has to be uniform in order to provide comfortable conditions. At the same time, illumination levels that are too high result in glare and visual discomfort. Uncontrolled direct solar illumination also affects the occupant's thermal comfort, since the infrared component of the daylight increases the interior thermal load (Herzog et al. 2004). Different approaches are thus required to redirect, diffuse, or limit penetration of daylight. Prolonged exposure to long-wave components of daylight may also affect surfaces of bio-based materials or interact with applied coating causing discoloration.

Different methods are available to redirect the light at the façade with some of these using bio-based materials. Standard aluminium slats in Venetian blinds could be replaced with wooden ones, having different reflection, absorption, and transmission characteristics. The fibrous structure of many bio-based materials could be used for diffusing the light if the used layer is sufficiently thin for the transmission. In Japan, there is an old tradition of making translucent paper using mulberry bark. The paper is glued to the lattice wooden frame to produce wooden sliding doors called shoji. The doors are commonly used to separate the interior from the exterior and to diffuse the light. Fibrous materials are also used in large-scale glazing, for example, in the factory Wilkhahn designed by Thomas Herzog (Dawson 2016).

Comfortable Sound Level (Acoustic Insulation)

To ensure suitable levels of sound, dwellings or offices must possess proper acoustic qualities which are related to sound transmission, absorption, and reflection. Sound transmission can occur via air (airborne sound transmission) or through a structure (structure-borne sound transmission). Different bio-based materials provide distinct sound propagation/insulation properties that vary with the environmental conditions. Acoustic properties of wood depend on its moisture content and grain direction. Density of timber also highly affects its acoustic properties due to the related differences in the porous/cellular internal structure. The most effective sound insulation in timber structures can be achieved by implementing the multi-skin concept and optimally arrange different wall layers (Herzog et al. 2012).

3.2 Aesthetics

The perception of aesthetical quality and related awareness of "beauty" changed over ages and will continue to change in future. Since it is challenging to study aesthetically pleasant qualities, it is difficult to provide definite guidelines (Eekhout 2008). Nevertheless, some universal attributes that are perceived as attractive for the built environment exist; therefore, it is possible to indicate at least a few important aesthetic guidelines for the application of building materials.

3.2.1 Aesthetical Measures

In general, it is difficult to explain what makes a quality aesthetically pleasing. It does not help that the aesthetical appeal of different qualities varies among individuals and often includes an emotional component. Aesthetics in qualities of façades are associated with prestige and symbolism and are meant to convey power and importance. A set of basic guidelines developed in classical architecture provides the basis for aesthetic assessment of buildings:

Symmetry

Symmetry is a sense of balance, defined as a state of equilibrium of the visual weights in a composition (Leopold 2006). Symmetry (bilateral/linear and radial) and asymmetry are both equally important design tools. Symmetry is usually linked with order, formality, and prestige and is historically associated with public or juridical buildings (Fig. 3.3). The opposite of symmetry is asymmetry, characterized by imbalance and disorder. While symmetry in some cases risks becomes too predictable, asymmetry can include complexity that conveys emotion and might be highly interesting.

Rhythm

Rhythm in design is defined as a regular and harmonious repetition of specific patterns (Leopold 2006). Rhythm could be repetitive or progressive (see Fig. 3.4). In the first case, rhythm is associated with the recurrence of forms (elements, colours), while in the latter rhythm is linked with the change of one characteristic of a motif (e.g., scale, colour).

Hierarchy

An example of rhythmical iteration of balconies on the timber façade hierarchy in design is observed when one element is emphasized more than others (Fig. 3.5). This can be manipulated by changing size, shape, and placement/orientation of building elements. Humans associate size with status; thus, the bigger element is

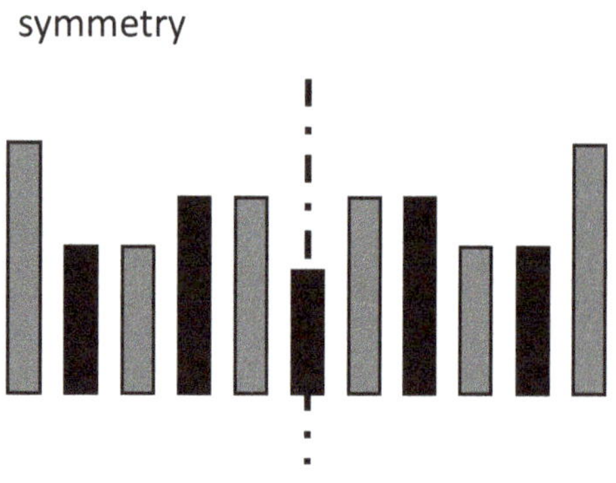

Fig. 3.3 Symmetry in architecture and its implementation

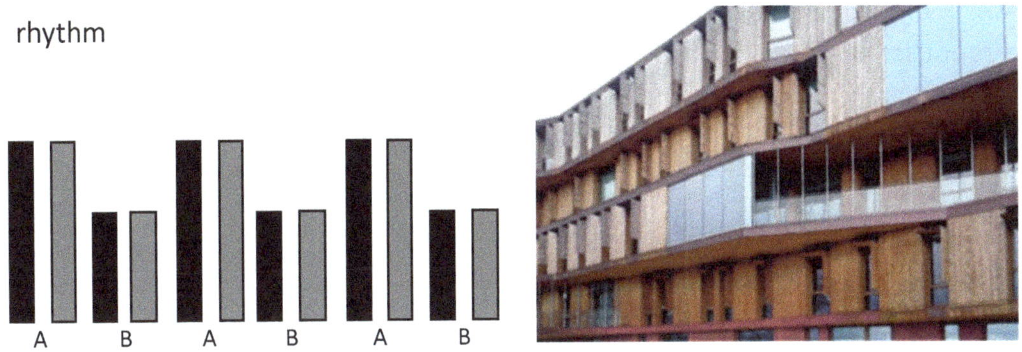

Fig. 3.4 Rhythm in architecture and its implementation. Image courtesy of Iztok Šušteršič—InnoRenew CoE

Fig. 3.5 Hierarchy in architecture and its implementation

perceived as more significant than the smaller one. In a group of identical elements, the one that is different in shape or form from the others will stand out. Similarly, in a group of identical items, the one that is placed in the centre will be perceived as more important even if it is equal to others in size. Hierarchy could also be discussed in the context of accentuation: when the relation of different design elements is described as balanced (almost of equal status), dominant (a case of hierarchical arrangement), or subordinated (when an element is visually less emphasized than others).

Proportion

Proportions are the relations between different dimensional elements of a form: lengths, areas, or volumes (Fig. 3.6). According to Euclid, a ratio refers to the quantitative comparison of two similar things, while proportion refers to the equality of ratios (Leopold 2006). Thus, a proportioning system establishes a consistent set of visual relationships between the parts of a building, as well as between the building elements and entire structure.

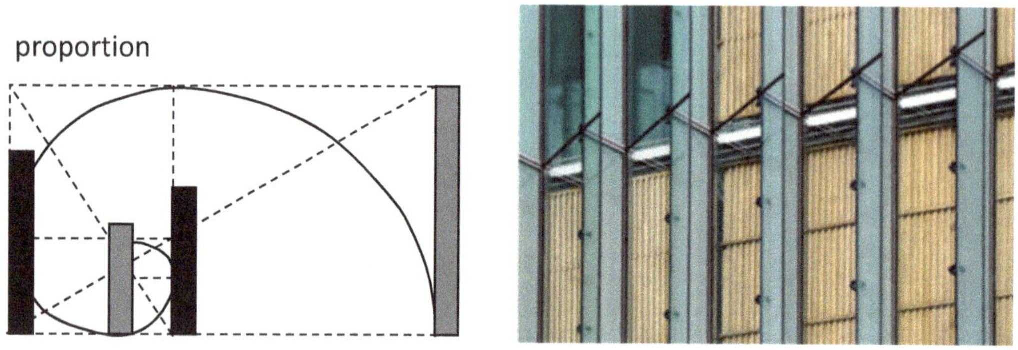

Fig. 3.6 Proportions in architecture and its implementation

Harmony

Harmony includes producing a visually pleasing relation between elements. Those elements usually share a common feature, trait, form, or colour (Fig. 3.7). Harmony in architecture is strongly associated with the compatibility with building surroundings (Salingaros 2017). While harmony is usually associated with the elements of higher order, the contrast is the juxtaposition of opposing elements (different materials, colours, and textures).

3.2.2 Surface Properties

Just like façades created from other building materials, bio-based façades are linked to the general aesthetical principles discussed above. Therefore, it is equally possible to strive towards symmetry or harmony by using timber, stone, or brick. However, there are important differences in surface quality between materials. Due to the specificity of bio-based materials, their surfaces are not uniform but contain diverse patterns. On a larger scale, these are related to the cladding direction (horizontal/vertical and plane/grilled), while on a smaller scale they are associated

Fig. 3.7 Harmony in architecture and its implementation—the Rotho Blaas SRL company, Cortaccia (BZ), Italy. Photograph courtesy of Rotho Blaas SRL

Fig. 3.8 Diagonal cladding on the expo building of Slovenia providing dynamic look. Image courtesy of SoNo arhitekti

with the wood grain direction (in case of timber façades). The pattern direction plays an important role in the visual perception of façades and the entire building. Vertical divisions will increase the perceived building height, while horizontal divisions will "visually" lower it and give the impression of denseness. Diagonal patterns typically imply a dynamic movement, transformation, and freedom (Fig. 3.8). If appropriately used they bring life, add volume, and make space feel larger than it is. However, if applied incorrectly, diagonal lines can increase a sense of confusion and imbalance. The surface heterogeneity and its potential creative use distinguish biofaçades from façades built from other materials.

Colour

In general, bio-based materials are characterized by warm, yellow, and brown hues. Wood-based products include a wide palette of colours commonly found in the nature, including pale yellow, orange, and brown. Cork, depending on the type, can be light yellow, brown, or even light blue. Deep natural-looking blue colouring usually results from the blue stain (sap stain) fungal infection and is considered to be a wood defect (Garau and Bruno 1993).

From an aesthetic perspective, uneven discoloration is a factor most damaging to the façade appearance. Discoloration is a process of gradual loss of the original colour. In bio-based products, it may appear as yellowing, browning, or greying. This process is also called bleaching when the original colours fade to grey. Non-protected wood exposed to UV radiation and moisture will change its colour (hue) depending on the cumulative dose of the weathering. The dose may vary along the façade surface resulting in uneven discoloration. Diverse wood species react to the weathering in different ways. Not only wood, but other bio-based materials also tend to change their colour under the influence of weather conditions. Expanded agglomerated cork, for example, weathers fast and uneven. The surface of the agglomerate becomes lighter, in contrast to the darker brown stains caused by

Fig. 3.9 Discolorations of bio-based building façades, with appreciated (**a**, **b**) and undesired (**c**, **d**) appearance

the manufacturing process (Roseta and Santos 2015). The discoloration (greying) of wood is often desired, especially when it is intended to emphasize the age of a structure (Brookes and Meijs 2008). Examples of building façade discoloration that might be perceived as positive and unpleasant are presented in Figs. 3.9a, b and 3.9c, d, respectively.

From a designer's perspective, the use of bio-based materials is always an asset, as bio-based materials are usually perceived as natural and user-friendly. Human mind pleasantly evaluates colours that are typical in biomaterials. Yellow, orange, and brown (especially the dark tones) are associated with the perception of physical comfort, warmth, security, and seriousness. The use of colours in façades can be understood through the colour use theory which is popular in diverse fields, such as visual composition in art (Garau and Bruno 1993). The simplest analogy concerning the colour use is the comparison of the art composition to music. As the colours of bio-based materials are rarely primary (yellow, blue, or red), they usually present a mix of so-called tertiary colours that are composed of the primary and secondary colours (green, violet, and orange). Colour research experts witness two types of reaction caused by colour: tension (leading to connection or separation) and attraction (leading to desirability or antipathy). Tensions originate from the lack of balance, due to maximal contrast or strong demand for chromatic completion, while attractions result from the presence or the absence of the primary colours. These guidelines should be used to optimally design façades and subsequently to select their colour (Garau and Bruno 1993).

Texture/Grain

From the aesthetic perspective, tangible/sensory qualities of biomaterials are generally rated high by users. Timber is appreciated as being warm to the touch thanks to its relatively high specific heat and the cellular structure. Human sense of touch is sensitive enough to detect the texture of the thin veneer itself. It can be easily identified in blind trials when compared to other materials. These qualities are transferred to many industrially processed wood-based products like chipboards (e.g., OSB with palpable texture) and particle boards (palpable only when the raw board is not covered with veneer or laminate). Other bio-based façade products also prove to be very attractive to users, with the cork as an example that is currently developed/researched to be used as an external cladding product. Textural properties of biomaterials are usually utilized in interiors, with multiple options possible regarding wood species and finishing options.

Textural properties of external timber cladding are perceived only in direct contact with the façade. However, it must be pointed out that timber elements are used in cladding at different heights; thus, in some cases, the direct access to them is limited. Exposure of biomaterials to external conditions facilitates natural ageing processes. In consequence, in many bio-based materials, the transformation becomes perceptible not only on a visual level (e.g., as discoloration) but also in others. Initially, flat-planed timber laths warp and split exposing texture of wood. Particularly, spring/winter grains of softwood exposed to weather for a prolonged period deteriorate at a varying speed/rate, thus making the wood texture more apparent. The same process occurs in other biomaterials, such as bamboo. The process of corrosion in biomaterials can be assessed from different perspectives. From the viewpoint of material durability, the surface directly exposed to the weathering is usually corroded the most.

Natural Roughness and Decorative Sculpture

Timber façade cladding can be produced as smooth (planed) but could also be deliberately sculpted. Wood carving is one of the world's oldest decorative techniques present almost in all cultures and regions (Figs. 3.10 and 3.11). This is due to the fact that wood is a relatively soft material that can be easily shaped with metal tools. Softwoods are especially simple to carve but more prone to the weather corrosion, as the water ingress is facilitated by the cross-cut grain. In general, wooden carved elements provide exceptionally tangible qualities, although this varies depending on the relief applied.

Fig. 3.10 Oldest wooden building in Mâcon (France), dated from the late fifteenth to early sixteenth century—La Maison du Bois, Mâcon, France. Image courtesy of Wim Willems—FirmoLin

Fig. 3.11 Sculptured façades of ancient Japanese buildings

3.2.3 Change in Appearance During the Service Life

Environmental factors (e.g., exposure to UV, rain and water condensation, temperature fluctuations, insect ingress) change most aspects of bio-based materials; however, material surfaces tend to be affected the most. Massive elements with smaller external surfaces are robust, while the slender ones are prone to faster degradation due to their relatively large surface area in relation to element's volume and mass. In general, the larger the external surface, the faster the degradation processes will occur. However, in bio-based façade claddings, the progress of ageing does not depend only on the S/V (surface/volume) factor but is also affected by a number of single-element surfaces that are exposed to the environmental conditions. A massive element with a single side exposed will weather/corrode at a lower rate than the slender element exposed to corrosion at all sides (Fig. 3.12). A detailed description of degradation processes during the service life of a building is provided in Chap. 5.

According to the number of sides of bio-based elements that are exposed to the external environmental conditions, façades can be divided into four typologies: not exposure, single-side exposure, double-side exposure, and the whole surface exposure.

Fully covered (unexposed) biomaterials are frequently used in cases where their exposure to external conditions can be severely destructive to the structure of the building's skin. The emerging façade typology in which timber constituting a façade skeleton or cladding is protected externally by a layer of flat glass is gaining acceptance. This solution combines two materials: renewable biodegradable timber and durable recyclable glass with isotropic physical and chemical parameters. This solution was originally introduced to visually enrich the spatial depth of the façade and reduce the heat loss but gradually evolved as a separate engineering solution and has been recently found as beneficial in the context of extending the service life

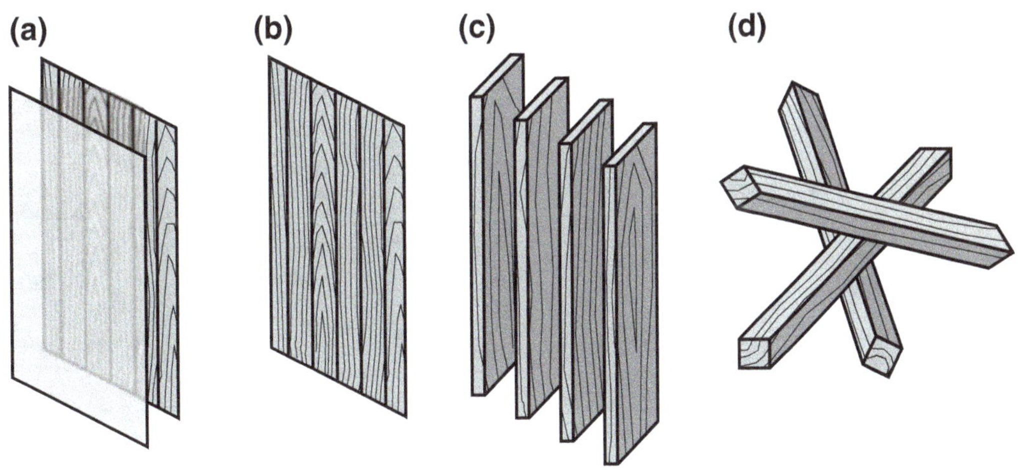

Fig. 3.12 Typology of bio-based building façades in regard to exposure: **a** not exposed, **b** single side, **c** double side, and **d** all surfaces exposed

of bio-based façades. Bio-based façades fully enveloped and sealed by glass are found mostly in structural applications. This typology was introduced in architecture by Japanese architect Shigeru Ban and used for the first time in GC Building in Osaka (arch. Shigeru Ban Architects, 2000), where the laminated timber structure was covered with a glazed façade (Fig. 3.13).

Fig. 3.13 Bio-based façade elements protected by external layer of glass, as implemented in CG Osaka Building (arch. Shigeru Ban Architects, 2000)

In opposition to fully sealed glass façades, a ventilated solution was independently developed in other buildings. Ventilated solutions are expected to perform better in terms of potentially destructive moisture capture, as the circulating air promotes drying of the surfaces. Timber in such façades is not fully protected from the environmental influence (it is susceptible to changes in temperature and humidity) but the rainscreen formed by the glass substantially affects the durability of the façade's cladding, as wetting is prevented. This façade typology is currently developing rapidly, as the issues of sustainable material use are gaining attention of investors and clients. An example of the double-skin façade is the FH 1 building at the campus of Frankfurt University of Applied Sciences (arch. Heribert Gies Architekten, MainzVoigt & Herzig Architekten & Ingenieure, 2007), where the external façade layer protects the timber cladding, while providing the space for the air exchange (Fig. 3.14). The external glass layer was also used in Market Hall in Ghent (arch. Marie-José Van Hee + Robbrecht & Daem, 2012), where the glass cladding protects the timber tiles used on the inclined roof surfaces and on vertical façade (Fig. 3.15).

An alternative approach for the double-skin façade is external layer of decorative wooden elements providing additional shadowing of building interior through the glass wall (Fig. 3.16).

Fig. 3.14 Double-skin façade with bio-based aesthetical components protected by glass and ventilation layers—FH 1 building at the campus of Frankfurt University of Applied Sciences (arch. Heribert Gies Architekten, MainzVoigt & Herzig Architekten & Ingenieure, 2007)

Fig. 3.15 Bio-based façade elements protected by external layer of glass, as implemented in Market Hall in Ghent (arch. Marie-José Van Hee + Robbrecht & Daem, 2012)

Fig. 3.16 Inversed double-skin façade in Cluny (France). Image courtesy of Wim Willems—FirmoLin

Single-side exposure of bio-based façade elements is widespread, as this type of exposure is typical for cladding. In such an arrangement, timber (or other materials) faces external environment with only one surface. These form large surfaces consisting of bio-based materials, usually containing a combination of smaller elements (boards or planks). Cladding exposed on a single side can be used in various orientations. The most popular is the horizontal orientation, as it enables an easy partial replacement of cladding elements, especially those that are located closer to the ground. The vertical orientation changes the proportion of façade, so the buildings seem to be thicker or longer than they really are. For both types, some extra space for air circulation has to be provided behind the cladding. In Hollainhof

Fig. 3.17 Single-side exposed bio-based façade as implemented in Hollainhof housing development in Ghent (arch. Neutelings Riedijk, 1998)

in Ghent (arch. Neutelings Riedijk, 1998), the horizontal cladding visually emphasizes the elongated building proportions (Fig. 3.17). On the contrary, in Wälderhaus in Hamburg (arch. Studio Andreas Heller GmbH Architects & Designers, 2012), the horizontal cladding orientation influences building perception and makes it visibly less dynamic, despite the overall faceted style.

Vertical cladding orientation is less frequent; however, it possesses an interesting potential in building façades. Such cladding orientations usually make buildings appear higher than they really are. This cladding type requires the horizontal substructure for the assembly process. Usually, this is provided by a system of laths. To ensure proper air circulation, laths must be positioned in a manner to allow air circulating upwards. The Westside Shopping and Leisure Centre in Bern-Brunnen (arch. Daniel Libeskind, 2008) is an example illustrating an optimal approach for implementing vertical bio-based cladding (Fig. 3.18).

Cladding positioned in arbitrary directions is rather rare as it is difficult to design them and to assemble. However, Yokohama Ferry Terminal (arch. Foreign Office Architects, 2003) represents an interesting example of a bio-based cladding used in the form of an undulating polygonized surface that smoothly transforms from the deck to the wall/façade (Fig. 3.19). The use of timber (teak) cladding is often found in ship deck flooring. Although such decks are visually impressive, their construction is technically challenging. Timber serves as a cover only, and the water is channelled in the membrane-covered layer below.

In double-side exposure, two main element surfaces are exposed to the external environment. This includes narrow and slender elements that are arranged perpendicularly to the façade's plane: vertical sun blinds, timber fins, and structural elements. Elements exposed on two sides visually enrich façades and create an impression of spaciousness and depth. In vertical element orientations, two main surfaces are exposed to different degrees which results in uneven ageing processes.

Fig. 3.18 Vertical orientation of cladding in Westside Shopping and Leisure Centre in Bern-Brunnen (arch. Daniel Libeskind, 2008)

Fig. 3.19 Dynamic transformation of the wall façade to the deck—Yokohama Ferry Terminal (arch. Foreign Office Architects, 2003)

Double-side exposure characterizes many finished buildings, especially those with façades containing timber fins or vertical sunshades. An interesting example is Asakusa Culture Tourist Information Center (arch. Kengo Kuma & Associates, 2013) that features vertical timber fins and internal louvers. This rhythmical and

(a) (b)

Fig. 3.20 Double-side exposure of bio-based building façade—**a** Asakusa Culture Tourist Information Center (arch. Kengo Kuma & Associates, 2013), **b** coffee shop in Delft

unique appearance leads to a unique building perception depending on the observer's standpoint. If viewed frontally, the façade seems to be made of glass, and if viewed diagonally, the building seems to be made of timber (Fig. 3.20a). Another example is a Sevenhills coffee shop in Delft. The glass façade is surrounded by vertical wooden elements creating external layer on entire façade (including roof) (Fig. 3.20b).

Whole surface exposure refers to elements that are fully exposed to environmental factors. This includes structural elements that are completely uncovered and left to the influence of the outside environment. Several such cases can be found in contemporary cities, also in the form of artistic installations. The example is the Pavilion of Reflections (arch. Studio Tom Emerson, 2016) that was temporarily erected on the Zurich Lake (Fig. 3.21). The large lattice structure is formed to house

Fig. 3.21 Pavilion of Reflections (arch. Studio Tom Emerson, 2016)

Fig. 3.22 Façade of Sunny Hills confectionery shop at Minami-Aoyama in Tokyo (arch. Kengo Kuma & Associates, 2013)

various cultural activities, including the open cinema. Roofs on this structure are supported by the spatial grid made of timber. The elements are fully exposed to the environmental influence. A different approach was adopted in Sunny Hills confectionery shop at Minami-Aoyama in Tokyo (arch. Kengo Kuma & Associates, 2013). A highly irregular grid of timber elements is shaped as a frayed cloud covering the building. Timber laths are positioned diagonally and interlock with each other at unusual angles. The structure gives the impression of being arbitrary and chaotic Fig. 3.22.

3.3 Safety

Façades contribute not only to building aesthetics but also to its technical performance, including its protection from the harsh environmental conditions. There are several safety-related issues associated with risks involved in design, construction, maintenance, repair, and overall performance of façades during their service life. The analysis of façade risks and failures is complex, since there are many materials involved, each with unique performance properties. Risk factors can be related to design, construction quality, or maintenance quality.

Aggressive environments can deteriorate and decrease the integrity and performance of façades. Therefore, façade components and the entire wall system should be primarily focused on protection from the weather. Humidity and water infiltration are the main causes of aesthetic deterioration in façades, since they lead to cracks and corrosion of the joints. The façade integrity can be improved by reducing the number of required joints by maximizing the grid size

Table 3.1 Safety aspects and mitigation action related to façade safety

Safety aspect	Mitigation action
Fire resistance	Using fire-resistant materials or materials treated with fire retardants Proper design that avoids fire spread
VOC emission (particularly regarding odour)	Using materials with particular VOC emissions
Falling façade elements (e.g., during hurricanes)	Proper mounting of façade elements, regular inspection, and maintenance
Resin leakage	Using modified wood or materials treated with coating system preventing resin leakage
Danger of human intrusion	Proper design, avoiding the use of horizontal elements
Dazzle (sunlight reflection)	Using matt elements, reorientation of elements to avoid reflection, external shading, planning of the nearby trees
Leaching of chemicals	Using materials resistant to percolating
Explosion resistance (shingles risks)	Using resistant materials, proper building façade design

(Moghtadernejad and Mirza 2014). Due to significant volume changes related to the shrinkage and swelling, façades from bio-based materials need to be designed with enough extra space allowing for dimensional changes. In addition, it should be taken into consideration that different environmental conditions related to specific building location (e.g., climatic, urban, rural zones) influence façade material deterioration uniquely.

High construction quality, including correct installation, high standards of workmanship, and quality control at the construction site, is necessary for assuring high building performance in future. Using modular structures and prefabrications (both are possible when using bio-based building materials) usually leads to fewer construction errors. During the building erection phase, it is important to consider precise and clear specifications allowing quality control of the construction progress on site. Systematic cleaning and inspections should be performed regularly to assure the desired aesthetic qualities and functional performance. Material selection and construction planning should be carefully carried out at the design phase to minimize future maintenance efforts. Moreover, it should be considered that a difficult access to façades results in costlier and riskier maintenance (Moghtadernejad and Mirza 2014). Other risks that are related to façade safety are summarized in Table 3.1. Some of the safety aspects listed are specifically concerning bio-based façades.

3.3.1 *Fire Safety*

Fire safety is the most important safety aspect related to the timber use in buildings. Causes of fires are almost never related to the material choice; thus, timber buildings do not ignite more often than buildings made of another material (Herzog et al. 2012). However, when the fire develops, timber behaves differently than other non-combustible materials. Since it can be considered a fuel, it is clear that its rapid oxidation during fire must always be taken into account. This results in large emissions of heat and smoke, both being potentially lethal to humans. Research shows the exposure to dangerous smoke fumes presents a greater threat to the inhabitants than a direct exposure to heat. In 80% of cases, the poisoning by toxic combustion products is the main cause for fire-related fatalities (Aseeva et al. 2014). Clearly, the smouldering of timber is a key factor, as during the process of combustion some of the most dangerous gases are produced (e.g., carbon monoxide).

When timber is used in buildings, it is essential to ensure adequate levels of fire protection. This can be done by preventing the ignition (e.g., by cladding timber in other non-combustible materials or using flame-retardant coatings) or by slowing down the combustion process (e.g., by using massive timber elements). Safety cladding is usually accomplished by using multi-layered gypsum fire-resistant plasterboard (approximately 1.2–1.5 cm thick) that covers the timber element. This method, although effective, has evident disadvantages. First, it requires work precision and is thus labour expensive. Second, it includes covering the timber, which may decrease the aesthetical appeal of buildings. Timber can be coated with fire-retardant chemicals that affect the smoke generation and decrease optical density of smoke during combustion. The chemicals also impede the timber ignition and increase the combustion temperature. Thus, coated timber can withstand in the fire longer since it needs higher temperatures to ignite. Timber coupled with fire retardants is classified as a moderately hazardous material at low heat flow intensity (Aseeva et al. 2014).

In buildings, massive large-sized glulam structural members (columns, beams, arcs, frames, trusses) are frequently used to bear high dead loads. Wood fulfils its structural function even at high temperatures in contrast to aluminium and steel, which lose their strength and rigidity at relatively low temperatures. Depending on the density, wood burns at the rate of approximately 40 mm per hour (Brookes and Meijs 2008). A charred layer appears on the wood surface and acts as a layer of insulation that prevents a temperature rise in the wood underneath, extending the load-bearing capability of timber members (Fig. 3.23). In massive timber structures, it is commonplace to calculate the necessary load-bearing section and then increase the cross section taking into account the intended time of fire resistance (e.g., 0,7 mm of section per minute of fire for softwood glulam) Recently, massive glulam walls and slab elements have become popular as sustainable alternatives to concrete. In massive elements, timber load-bearing capacity in a fire increases with larger dimensions; thus, glulam is hard to ignite and burns slowly (Brandon and Östman 2016).

Bio-based façades pose a unique issue that needs to be considered to fulfil the fire safety requirements. Façade cladding is typically installed in the so-called ventilated façade system. Here, external cladding is not mounted directly on the wall structure, so a layer of an air plenum is present behind the cladding. Air circulates in the plenum to provide the vapour and moisture exchange. The wall

Fig. 3.23 Carbonized surface of the CLT panel after fire test preserving structural integrity of the wall

component beneath (usually insulation) is sealed with vapour permeable membranes. In an event of a fire, air present in the plenum facilitates the development of the fire. When the temperature increases, the chimney effect behind the cladding is observed, which leads to a flame spreading towards the top of the façade (Jeffs et al. 1986). Proper design of horizontal and vertical fire stops, or proper compartmentation of the entire façade's surface may prevent fire distribution. If properly designed, a timber-clad façade can meet the requirements of REI60 (60 min of adequate performance in load-bearing capacity, integrity, and insulation) for a medium fire duration (Lenonn 2008).

3.4 Adaptation and Special Functionalities

Traditional building envelopes act as a barrier between the indoors and the outdoors. The main developments in design focus on structure and energy aspects. The trend for next-generation façades—called adaptive, dynamic, or responsive—is gradually recognized and desired (see Chap. 1 for more details). Key emphases in new façade systems are shifting:

- From the barrier to the interface
- From invariable and static to responsive and dynamic
- From passive and single functional to adaptive and multifunctional
- From conventional to customized.

In consequence, innovative façades are able to modify its structure, function, and behaviour and are responsive to change in harsh environmental conditions. Their overall goal is to improve building performance.

3.4.1 *Self-adaptive Biomaterials*

Inspiration in novel design strategies is frequently adopted from the field of biology, namely, from the processes of organism adaptation to diverse climate conditions. Bio-based materials have a high potential to be used in adaptive façades, far beyond the use of glass-framed timber elements or the exchange of non-bio-based to bio-based components (Callegaria et al. 2015). Bio-based materials are characterized by inherent properties that change according to variations in the surrounding environmental conditions. Such changes are autonomous and do not require any control mechanisms. To take advantage of these mechanisms, structure design and material engineering must be carried out properly.

Hygroscopic Devices/Energy Control

Cellulose, as the major constituent of most biomaterials, is an example of an adaptive natural material, as it changes its dimensions according to variations in relative humidity and corresponding moisture content equilibrium (Hill and Xie 2012). If shaped properly, cellulose-based elements can be used as actuators facilitating rotating, skewing, or bending façade elements. Such a functionality is so-called passive humidity-driven actuation. Max Planck Institute of Colloids and Interfaces in Potsdam (Germany) manages a long-term research programme on "Biomimetic Actuation and Tissue Growth". It includes both the distortion of thin layers of biomaterials with different properties and the actuation provided by pressurized anisotropic honeycomb structures (e.g., in D. nakurense seed capsules). Two veneer-laminated wood layers with different properties bend depending on the humidity levels due to the different expansion coefficients and physical properties of wood (related to its grain direction). This absorption mechanism was used in many conceptual designs, and some of them reached a prototype stage. Architect Achim Menges proposed a "Hygroscope", where dimensional instability of wood is used as an object responsive to climate changes that is capable to react to moisture content without external control (Menges 2012). The triangular flakes that were developed by Menges are constructed from laminated quarter-cut maple veneer. This method of cutting bisects annual growth rings that not only results in a straight grain or ribbon-striped appearance but also facilitates the distortion of the flakes. Similar shape change mechanism could also be employed as a building-integrated photovoltaics (BIPV) orientation device depicted in research done by Mazzucchelli and Doniacovo (2017). In this project, thin-film solar cells were coupled with a thin layer of hydromorphic material capable of responding to changes in environmental humidity by modifying its own curvature (Fig. 3.24).

Bio-based Latent Heat Storage/Energy Conservation

Latent heat storage is one of the most efficient ways of storing thermal energy. As such, it is actively changing the properties of adaptive façades and reducing their energy use. Here, the most common crude oil-based paraffin and salt hydrates are regarded as the optimal solutions. Phase change material (PCM) should be handled

(a)

(b)

(c)

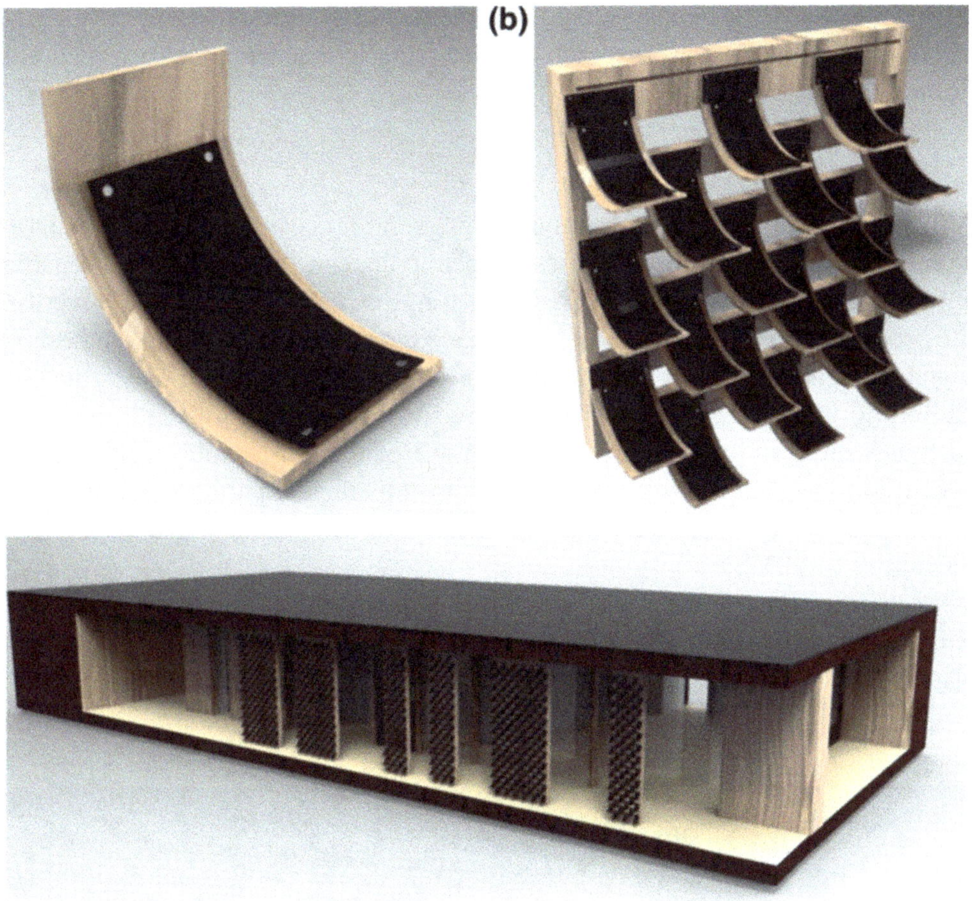

Fig. 3.24 Bio-based materials implemented as building-integrated photovoltaics: **a** single active element, **b** one module of the active façade, **c** integration of the façade with a modern building. Image courtesy of Enrico Sergio Mazzucchelli—Politecnico di Milano

with care, as most of them are highly flammable or/and toxic. One of the ways of reducing potential dangers is to use the method of encapsulation, especially microencapsulation that enables a direct use of PDM (e.g., in concrete). Here, some bio-based materials present many application possibilities and, accordingly, are widely recognized as bio-based phase change materials. Usually, a mix of paraffinic and bio-based compounds or blend of fatty acids is prepared to change from liquid to solid and vice versa at a given temperature (Mathis et al. 2018). One of the most recent solutions is BioPCM™ material marketed in Australia as "room temperature ice" (Ramakrishnan et al. 2017). BioPCM™ absorbs excess heat during the day and releases it back in the evening as buildings cool. The material is designed to be installed in ceilings and upper parts of the walls to stabilize diurnal temperature changes (it starts melting at the temperature of 20 °C and becomes fluid at the temperature of 24 °C). The product uses waste products from the food manufacturing process. Although BioPCM™ is biodegradable, it cannot serve as a food source and is thus not targeted by animals. The material is delivered in the form of a rolled matt with PCM encapsulated in the cells of a flame-retardant poly film. It is installed as an internal layer to act as a heat sink in the extreme Australian climate (Ramakrishnan et al. 2017).

Energy Generation (Biomass)

Although they are deriving from living things, no processes of life are taking place within biomaterials. If life processes happen to appear, they are usually regarded as undesired or even dangerous (e.g., when fungi or moulds develop). Nevertheless, life processes based on energy and substance circulation (e.g., photosynthesis) pose a great potential in energy generation, as the solar energy could be transformed directly to biomass. Apart from being used for energetic purposes, some species, such as willow and poplar, are becoming one of the most widely cultivated plants because of their high fertility and yield as well as their capabilities of phytomining and phytoremediation of soils contaminated with heavy metals (Tlustoš et al. 2007; Sandak et al. 2017).

Prototypes representing innovative and transparent energy-harvesting systems were developed at the IBA Hamburg in 2013 in the BIQ housing building (called also Algae House) (Fig. 3.25). The developed system features a thin glazed tank

Fig. 3.25 BIQ (arch. SPLITTERWERK, Arup GmbH, B+G Engineers, Immosolar, 2013)

called "bioreactor" which creates a biohabitat for algae. The tanks containing a nutrition solution are located on the south-facing façade of the building. A separate water circuit running through the façade continuously supplies the algae with liquid nutrients and carbon dioxide. With the aid of sunlight, algae can photosynthesize and grow. Algae were chosen due to their efficiency in producing biomass. The developer explained that algae produce up to five times as much biomass per hectare than other biomass production systems. Moreover, algae growing in a building in Hamburg contain a high proportion of oils that can be directly used to generate energy (Elrayies 2018).

3.4.2 *Special Functionalities of Bio-based Façades*

Bio-based materials can be used to perform certain special functions. Microclimate regulation and protection from wind are among the conventional tasks of a building envelope, while some specially shaped bio-based elements can take the role of supplementary tasks and appear in shades, louvers, and shutters. Bio-based materials can be also used for decorative purposes.

Conventional building elements are typically made of timber that no longer contains any life processes. New possibilities arise if the use of living organisms in the façade envelope is considered. Algae can be used to create biomass that is later used in energy production, as previously described. Using live plants introduces numerous applications, including water retention, air filtering, wind gust protection, and heat gain reduction. Those functions are usually accomplished by greenery arranged in different forms in so-called vertical gardens that provide the framework for vertical rather than horizontal plant growth. In façades, the greenery layer is typically located at the blind (windowless) sections of walls (Fig. 3.26) but may be also deliberately positioned to the front of the window to provide shading and noise protection as well as the light-filtering function (Fig. 3.27).

Fig. 3.26 Green façade on blind sections of the wall

Fig. 3.27 Institute of Physics in Berlin-Adlershof (arch. Augustin Und Frank Architekten, 2003)

Vertical farming presents a particular challenge since plants providing food are typically grown isolated from the external environment, even when the daylight is used. This enables the regulation of heat and moisture flow and reduces the energy use while controlling pests. In many existing examples (a vertical farm in Japan is producing food on the industrial level), food is produced using artificial LED lighting that emits only the radiation wavelengths that are absorbed by plants during their growth (Goldstein 2018).

Urban pollution is another concern that can be mitigated by façade design. A new trend of smog-eating façades aims to improve air quality in highly populated cities. Recently developed façade materials aim to capture CO_2 from the air. One such material—"made of air" or "hexChar"—is composed of biochar (http://www.madeofair.com/). Designers aim to develop carbon-negative building materials, where material production and use phase lower environmental impacts of buildings. Note that this is opposite to the traditional building practice. Alternative approach for sequestering and storage of carbon might be inspired by biomimetic. There are several organisms and processes in nature that are able to store, sequester, or recycle carbon. Understanding mechanism of those processes might lead to the development of technologies suitable for industrial processes and the built environment (Zari 2015).

References

Aseeva R, Serkov B, Sivenkov A (2014) Fire behavior and fire protection in timber buildings. Springer, Wien

ASHRAE Standard 55-2004 (2004) Thermal environmental conditions for human occupancy. American National Standards Institute, Atlanta

Attias N, Danai O, Ezov N et al (2017) Developing novel applications of mycelium based biocomposite materials for design and architecture. In: Final COST FP1303 conference "Building with bio-based materials: Best practice and performance specification", Zagreb

Boswell K (2013) Exterior building enclosures: design process and composition for innovative façades. Wiley

Brandon D, Östman B (2016) Fire safety challenges of tall wood buildings. Fire Protection Research Foundation, USA

Brookes AJ, Meijs M (2008) Cladding of buildings. Taylor and Francis, Abingdon

Callegaria G, Spinelli A, Bianco L et al (2015) NATURWALL©—A solar timber façade system for building refurbishment: optimization process through in field measurements. Energy Procedia 78:291–296

Hill CAS, Xie Y (2012) The water vapour sorption properties of cellulose. In: Harding SE (ed) Stability of complex carbohydrate structures, biofuels, foods, vaccines and shipwrecks. Royal Society of Chemistry, pp 137–152

Dawson L (2016) Wilkhahn factory in Bad Münder, Germany by Thomas Herzog. Arch Rev

Eekhout M (2008) Methodology for product development in architecture. Delft University Press/ IOS Press, Delft

Elrayies GM (2018) Microalgae: prospects for greener future buildings. Renew Sustain Energy Rev 81:1175–1191

Frihart CR (2016) Potential for Biobased adhesives in wood bonding. In: Proceedings of the 59th international convention of society of wood science and technology, Curitiba, Brazil

Garau A, Bruno N (1993) Colour harmonies. University of Chicago Press

Goldstein H (2018) The green promise of vertical farms. IEEE Spectr 55(6):50–55

Herczeg M, McKinnon D, Milios L et al (2014) Resource efficiency in the building sector. Final report prepared for European Commission by ECORYS and Copenhagen Resource Institute, Rotterdam, Netherlands

Herzog T, Krippner R, Lang W (2004) Façade construction manual. Birkhäuser, Basel

Herzog T, Natterer J, Schweitzer R et al (2012) Timber construction manual. Birkhäuser, Basel

ICSC: 0275 (2012) Formaldehyde. The National Institute for Occupational Safety and Health

Ikei H, Song C, Miyazaki Y (2017) Physiological effects of wood on humans: a review. J Wood Sci 63:1–23

Jeffs GMF, Klingelhöfer HG, Prager FH et al (1986) Fire-performance of a ventilated façade insulated with a B2–classified rigid polyurethane foam. Fire Mater 10(2):79–89

Kim YS, Lee KH, Kim JSJ (2016) Weathering characteristics of bamboo (Phyllostachys puberscence) exposed to outdoors for one year. Wood Sci 62(4):332–338

Knaack U, Klein T, Bilow M et al (2007) Façades: the principles of construction. Birkhäuser, Basel

Lennon T (2008) The fire performance of timber façades test report number 243-505. BRE Certification Ltd

Leopold C (2006) Geometry concepts in architectural design. In: 12th international conference on geometry and graphics, Salvador, Brazil

Lovel J (2013) Building envelopes: an integrated approach. Princeton Architectural Press, New York

Mathis D, Blanchet P, Landry et al (2018) Thermal characterization of bio-based phase changing materials in decorative wood-based panels for thermal energy storage. Green Energy Env (in press). https://doi.org/10.1016/j.gee.2018.05.004

Mazzucchelli E, Doniacovo L (2017) The integration of BIPV adaptive flakes in the building envelope. In: 12th conference on advanced building skins, Bern, Switzerland
Menges A (2012) Material computation: higher integration in morphogenetic design. Wiley, London
Moghtadernejad S, Mirza SM (2014) Service life safety and reliability of building facades. In: Second international conference on vulnerability and risk analysis and management (ICVRAM) and the sixth international symposium on uncertainty, modeling, and analysis (ISUMA)
Noti JD, Blachere FM, McMillen CM et al (2013) High humidity leads to loss of infectious influenza virus from simulated coughs. PLoS ONE 8(2):e57485
Raja DS, Sultana B (2012) Potential health hazards for students exposed to formaldehyde in the gross anatomy laboratory. J Env Health 74:36–40
Ramakrishnan S, Wang X, Sanjayan J et al (2017) Thermal performance of buildings integrated with phase change materials to reduce heat stress risks during extreme heatwave events. Appl Energy 194:410–421
Roseta M, Santos CP (2015) Study in real conditions and in laboratory of the application of expanded agglomerated cork as exterior wall covering. Key Eng Mater 634:367–378
Salingaros NA (2017) Unified architectural theory: form, language, complexity. Sustasis Foundation Portland, Oregon, USA
Sandak A, Sandak J, Waliszewska B et al (2017) Selection of optimal conversion path for willow biomass assisted by NIR spectroscopy. iForest 10:506–514
Sid N (2018) Development and demonstration of highly insulating, construction materials from bio-derived aggregates. In: Proceedings of 17th edition of international conference on emerging trends in material science and nanotechnology, Rome, Italy
Stevenson F, Baborska-Narożny M (2018) Housing performance evaluation: challenges for international knowledge exchange. Build Res Inf 46(5):501–512
Tlustoš P, Száková J, Vysloužilová M et al (2007) Variation in the uptake of arsenic, cadmium, lead, and zinc by different species of willows Salix spp. grown in contaminated soils. Open Life Sci 2(2):254–275
Xing Y, Brewer M, El-Gharabawy H et al (2018) Growing and testing mycelium bricks as building insulation materials. IOP Conf Ser Earth Env Sci 121:(2018)
Zari MP (2015) Can biomimicry be a useful tool for design for climate change adaptation and mitigation? In: Torgal FP et al (eds) Biotechnologies and biomimetics for civil engineering. Springer International Publishing Switzerland

Chapter 4
Best Practices

Form must have a content, and that content must be linked with nature.

Alvar Aalto

Abstract This chapter presents examples of innovative and sustainable use of bio-based products in building façades. Selected cases represent best practices implemented all over the world are presented from the perspective of art and innovation. Cases and examples incorporated in this chapter are indexed and summarized in a form of portfolio. Short description regarding materials, building function as well as motivation of architects to design certain objects is presented. Buildings presented here may inspire new generation of architects to successfully implement biomaterials-based solutions in their future projects.

The design of a building is an important process affecting its operational performance. Sustainable, safe, and comfortable built environment is particularly essential in present day, when a majority of people spend most of their time in offices, factories, or homes. Buildings are gradually becoming people centric. The focus on the occupant well-being and comfort as well as on flexible, collaborative, and adaptable spaces has become a concern of architects, designers, and developers (Jadhav 2016). New materials, design tools, and building technologies encourage buildings to be more responsive to their occupants and the environment. By using renewable materials in buildings and adopting sustainable design practices, we can

A. Sandak et al., *Bio-based Building Skin*, Environmental Footprints and Eco-design of Products and Processes, https://doi.org/10.1007/978-981-13-3747-5_4

contribute to human health and well-being. Nowadays, contemporary architectural practice is driven by the innovation of materials. Advanced, smart, responsive, and biologically inspired materials are gaining popularity in architectural design (Aksamija 2016). Wood and other bio-based building materials have gradually become more important, especially in the context of CO_2 neutral economy. None of the other materials can be utilized in such numerous ways as wood, as it is remarkably versatile, aesthetically charming, and, at the same time, entirely recyclable. The examples presented in this chapter display the wealth of innovative and sustainable use of bio-based products in building façades. The products used include modified wood, engineered timber products, as well as certain exploratory solutions incorporating bio-based energy harvesting systems. The function of selected buildings was not limited to the physical space they offer.

We would like to present the following case studies (Figs. 4.1, 4.2, 4.3, 4.4, 4.5, 4.6, 4.7, 4.8, 4.9, 4.10, 4.11, 4.12, 4.13, 4.14, 4.15, 4.16, 4.17, 4.18, 4.19, 4.20, 4.21, 4.22, 4.23, and 4.24) with the aim to demonstrate how buildings interact with the occupants and what motivated architects along the design process. The knowledge-based implementation of bio-based materials in architecture is restricted only by the imagination of designers. We sincerely hope that buildings presented here may inspire new generation of architects to successfully implement biomaterials-based solutions in their future projects.

01. Asakusa Culture Tourist Information Center

ARCHITECT:
Kengo Kuma & Associates

COMPLETION DATE:
2013

LOCATION:
Taito-ku, Tokyo (JP)

DESCRIPTION OF THE CASE-STUDY:
It's made mainly from glass and wood, and its interiors are airy and bright. The building benefits from lots of natural light, and excess sunshine is filtered by vertical wooden louvres that wrap the building´s eight stories.

REFERENCES:
https://inhabitat.com/kengo-kuma-designs-a-

Fig. 4.1 Asakusa Culture Tourist Information Center

02. Bibliothèque nationale de France

ARCHITECT:
Dominique Perrault

COMPLETION DATE:
1996

LOCATION:
Paris (F)

DESCRIPTION OF THE CASE-STUDY:
The double facade is visually complex and layered. The wooden screens protecting the books age and acquire a patina without loosing their qualities. Glass, steel and wood are combined here and provide sense of cohesion and individuality to the reading rooms.

REFERENCES:
http://www.bnf.fr/documents/dp_perrault_en.pdf

Fig. 4.2 Bibliothèque nationale de France

03. BIQ

ARCHITECT:
SPLITTERWERK, Arup GmbH, B+G Engineers, Immosolar

COMPLETION DATE:
2013

LOCATION:
Wilhelmsburg, Hamburg (DE)

DESCRIPTION OF THE CASE-STUDY:
At close range, the façades, oscillating from afar through the constantly growing algae, start to move. Bubbles forming through the supply of carbon dioxide and nitrogen, as well as the permanently essential circulation of water containing aerosol-like microalgae, seem to suggest that biomass production could be a solar-powered art installation, steadily bubbling along.

REFERENCES:
http://syndebio.com/biq-algae-house-splitterwerk/

Fig. 4.3 BIQ building

04. Cafe Birgitta

ARCHITECT:
Talli Architecture and Design

COMPLETION DATE:
2014

LOCATION:
Helsinki (FI)

DESCRIPTION OF THE CASE-STUDY:

Shou-sugi-ban, or burnt wood siding is a traditional Japanese method for charring cypress boards and to use them as cladding. It is known, to last over 80 years and is a natural protectant against fires and insects. The wood surface becomes water-proof through the carbonisation and is thus more durable. The process is quite simple and the result is bold and beautiful. In in Pyhä Birgitta Park in Helsinki cafe Birgitta with charred wood cladding is an example of architectural design that takes advantage of both traditional and new, innovative solutions.

REFERENCES:

https://binodem.wordpress.com/2015/02/17/wood-charring-machine/

Fig. 4.4 Cafe Birgitta

05. Chapelle de St-Loup

ARCHITECT:
Localarchitecture

COMPLETION DATE:
2008

LOCATION:
Pompaples (CH)

DESCRIPTION OF THE CASE-STUDY:
Localarchitecture, which has a special interest in timber construction and new structural solutions, is known of several works exploring traditional and contemporary wood building techniques. The team developed a structure using timber panels, which makes it possible to cover large areas with fine sections. The shape was generated using computer software that calculates the load-bearing structure, determines the dimensions and transmits this information to the machine that cuts out the 6-cm thick timber panels.

REFERENCES:
https://www.archdaily.com/9201/temporary-chapel-for-the-deaconesses-of-st-loup-localarchitecture

Fig. 4.5 Chapelle de St-Loup

06. European Council and Council of the EU

ARCHITECT:
SAMYN and PARTNERS

COMPLETION DATE:
2016

LOCATION:
Brussels (BE)

DESCRIPTION OF THE CASE-STUDY:
In the context of a sustainable development approach, it was decided to restore and reuse some old, though still efficient, window frames. Glazed double façade is made from an outer skin patchwork of recycled old oak windows found from demolition sites, with crystal clear single glazing, and an inner skin of crystal clear double glazing. The final design provides the necessary acoustic barrier from the traffic noise of Rue de la Loi in the European Quarter of Brussels and a first thermal insulation for the inner space.

REFERENCES:

https://www.designboom.com/architecture/philippe-samyn-and-partners-europa-building-eu-council-headquarters-belgium-04-05-2017/

Fig. 4.6 European Council and Council of the EU

07. Market Hall in Ghent

ARCHITECT:
Marie-José Van Hee
Robbrecht & Daem

COMPLETION DATE:
2012

LOCATION:
Ghent (BE)

DESCRIPTION OF THE CASE-STUDY:
The building is located in the centre of Ghent close to historic stone buildings. It is designed to encourage local residents to convene in the market square for social gatherings and public events. The building materials include concrete columns, a steel framed roof and a wooden cladding. Additionally, a glass envelope protects the wood and provides a soft shine, with the reflected sky.

REFERENCES:
https://www.dezeen.com/2013/04/19/market-hall-by-robbrecht-en-daem-architecten-and-marie-jose-van-hee-architecten/

Fig. 4.7 Market Hall in Ghent

08. Parkeergarage Laakhaven

ARCHITECT:
Moke Architects

COMPLETION DATE:
2014-2017

LOCATION:
Hague (NL)

DESCRIPTION OF THE CASE-STUDY:
For the first time in history bamboo poles were used for a parking garage siding in the Netherlands, which makes this project more environmentally friendly, sustainable and innovative. The entire facade is covered with the highest quality bamboo poles (2,370 lineal meters) imported from Colombia.

REFERENCES:

https://www.bambooimport.com/en/blog/bamboo-cladding-parking-garage-the-hague

https://www.mokearchitecten.nl/portfolio/parkeergarage-laakhaven/

Fig. 4.8 Parkeergarage Laakhaven

09. Pavillon of Reflections

ARCHITECT:
Studio Tom Emerson
Manifesta 11, ETH Zurich

COMPLETION DATE:
2006

LOCATION:
Zurich (CH)

DESCRIPTION OF THE CASE-STUDY:
A timber island, arrange like a fragment of intimate urban space enclosed by six objects: a tower, a tribune, a bar, a sun deck with changing cubicles below, a central pool with cinema screen above, and three generous sets of steps that lead into the lake. Together with the tower, the volumetric roofs over the bar are built up from a distinct profile of timber lattice roofs.

REFERENCES:
http://www.bnf.fr/documents/dp_perrault_en.pdf

Fig. 4.9 Pavilion of reflections

10. Sampa Pauma House 6

ARCHITECT:
KUS Corporation

COMPLETION DATE:
2013

LOCATION:
Setagaya-ku, Tokio (JP)

DESCRIPTION OF THE CASE-STUDY:
Wood elements are used here as elements providing frame that surrounds the building. By the use of timber as "tree grass lattice" at the outer circumference of the building, architects aimed to create attractive both interior and exterior spaces.

REFERENCES:
http://www.kus.co.jp/work-shimouma.html
https://www.early-age.co.jp/detail/?sanpapa_shimouma_house

Fig. 4.10 Sampa Pauma House 6

11. SunnyHills at Minami-Aoyama

ARCHITECT:
Kengo Kuma & Associates

COMPLETION DATE:
2013

LOCATION:
Minami-Aoyama, Tokyo (JP)

DESCRIPTION OF THE CASE-STUDY:
The intricate structure takes inspiration from traditional bamboo baskets, but the angle of the wooden lattice is very unique. Unlike the conventional 90 degrees, the slats are angled at 30 and 60 degrees assembled with „no-glue-nor-screws", a technique Kuma considers „the essence of Japanese architecture".

REFERENCES:
https://inhabitat.com/kengo-kuma-wraps-tokyos-sunny-hills-cake-shop-in-a-latticed-3d-wooden-cloud/

Fig. 4.11 SunnyHills at Minami-Aoyama

12. Tietgenkollegiet — Student Housing

ARCHITECT:
Lundgaard & Tranberg Architects

COMPLETION DATE:
2006

LOCATION:
Copenhagen (DK)

DESCRIPTION OF THE CASE-STUDY:
The facade is made with a unique copper alloy and features sliding partitions inspired by traditional southern Chinese Hakka house architecture. The alloy keeps the building surface clean and protected, and it will age to a rich dark tone over time, allaying restoration needs in the future. American oak and glass partitions alternate throughout the alloy wall, creating an exciting and dynamic facade that also encourages the flow of fresh air and sunlight.

REFERENCES:
https://inhabitat.com/lundgaard-and-tranbergs-tietgenkollegiet-dorm-is-the-coolest-circular-housing-on-campus/

Fig. 4.12 Tietgenkollegiet—student housing

13. Wälder Haus

ARCHITECT:
Andreas Heller Architects & Designers

COMPLETION DATE:
2014

LOCATION:
Helsinki (FI)

DESCRIPTION OF THE CASE-STUDY:

Wood play here an important role in two respects. It is used as a sustainable building material in the building structure and as façade material. Moreover the WÄLDERHAUS host an exhibition that deals with the relationship between the forest, the city and its inhabitants.

The upper three storeys are made entirely of larch wood. The green roof provides habitat for plants and animals and it can be colonized by birds and insects.

REFERENCES:

https://www.iba-hamburg.de/projekte/wilhelmsburg-mitte/eingangskomplex-am-inselpark/waelderhaus/projekt/waelderhaus.html

Fig. 4.13 Wälder Haus

14. Yokohama Ferry Terminal

ARCHITECT:
Foreign Office Architects

COMPLETION DATE:
2002

LOCATION:
Yokohama (JP)

DESCRIPTION OF THE CASE-STUDY:
Throughout the project, a deliberate dynamism pervades the tectonic and material languages of the building. The abundance of non-orthogonal walls, floors, and ceilings creates a controlled sense of vertigo that is accentuated by similarly off-kilter fixtures and details. The effect is magnified by material cues, such as the shifting grains of the wooden planks on the observation deck that indicate the locations of creases, and the minimalist grey metal paneling that is revealingly worn by the structures under it.

REFERENCES:
https://www.archdaily.com/554132/ad-classics-yokohama-international-passenger-terminal-foreign-office-architects-foa

Fig. 4.14 Yokohama Ferry Terminal

15. Kreod Pavillion

ARCHITECT:
Chun Qing Li

COMPLETION DATE:
2012

LOCATION:
London (UK)

DESCRIPTION OF THE CASE-STUDY:

KREOD is an innovative architectural sculpture, organic in form, environmentally-friendly and inspired by nature. Resembling three seeds, the three $20m^2$ pods combine through a series of interlocking hexagons to create an enclosed structure that is not only magnificently intricate but secure and weatherproof. KREOD functions beautifully both as an architectural landmark and an imaginative exhibition space – its three pods can be combined in a variety of configurations or installed as independent free-standing forms.

REFERENCES:

https://kebony.com/us/projects/kreod/

Fig. 4.15 Kreod pavilion. Photograph courtesy of Kebony

16. Rotho Blaas SRL headquarter's extension

ARCHITECT:
monovolume architecture + design

COMPLETION DATE:
2005

LOCATION:
Kurtatsch (IT)

DESCRIPTION OF THE CASE-STUDY:
The Rothoblaas office is a large scale commercial operation specializing in assembling systems and power tools for the woodworking industry. The aim of the project was to create a compact building with a high level of recognition. The building acts as corporate identity of the enterprise; contemporary, representative and innovative. This has led to a functional, compact structural shell, provided with a glass envelope. The main building material employed is wood, in order to show their own products.

REFERENCES:
https://www.archdaily.com/36359/rothoblaas-limited-company-monovolume?ad_medium=gallery

Fig. 4.16 Rotho Blaas SRL headquarters. Photograph courtesy of Paolo Grossi

17. Barangaroo House

ARCHITECT:
Collins and Turner

COMPLETION DATE:
2017

LOCATION:
Sydney Harbour (AU)

DESCRIPTION OF THE CASE-STUDY:

Barangaroo House, a free-standing, three-storey restaurant, has become one of the first projects in Sydney to utilise Accoya® wood cladding. The world-leading high performance, sustainable wood product, and the distinctive Japanese charring technique, Shou Sugi Ban has been merged here. „Utilising Shou Sugi Ban was an ideal way to create a unique, striking building form which references ancient craftsmanship and traditions in a very contemporary way"*

*Huw Turner

REFERENCES:

https://www.dezeen.com/2018/08/13/collins-and-turner-barangaroo-house-restaurant-sydney-architecture/

Fig. 4.17 Barangaroo house. Photograph courtesy of Rory Gardiner

18. 1500 West Alabama Office Building

ARCHITECT:
Dillon Kyle Architects

COMPLETION DATE:
2018

LOCATION:
Houston (US)

DESCRIPTION OF THE CASE-STUDY:
The exterior has a commercial boxy look, so the team looked for creative cladding finishes to wrap the building and provide a rainscreen. Accoya® wood boards covers the entire building. The rainscreen is made of G80 galvanized steel sheets offset from the building with the Accoya® leaf boards attached to it. An abstract leaf-like pattern is carved into 2,500 eight-foot long by eight-inches tall and 11/16-inches in depth. The leaf pattern serves as a gentle reference to the live oak trees that line the neighbourhood.

REFERENCES:

https://www.accoya.com/projects/project/dillon-kyle-architects-office-building-gives-the-wow-factor-with-accoya/

https://www.dkarc.com/dillon-kyle-architects-houston-tx.html

Fig. 4.18 1500 West Alabama office building. Photograph courtesy of Accsys

19. Misono Branch of the Hekikai Shinkin Bank

ARCHITECT:
Kengo Kuma

COMPLETION DATE:
2017

LOCATION:
Nagoya (JP)

DESCRIPTION OF THE CASE-STUDY:

The seven-story Misono branch building is located in the Naka Ward of Nagoya. The Japanese modern style design is highlighted by eye-catching Accoya® planks diagonally attached on the exterior glass walls and tall trees planted as a green-void corner feature near the base of the building. Constructed by Sekisui House, Ltd. Accoya® was selected due to its dimensional stability making it possible to achieve an elegant, slim structure having wide fixing points without risk of bending and warping.

REFERENCES:

https://www.accoya.com/projects/project/hekikai-shinkin-bank-chooses-accoya-wood-facades-asia/

https://www.designboom.com/architecture/kengo-kuma-accoya-wood-cladding-toki-shinkin-bank-misono-branch-07-06-2018/

Fig. 4.19 Misono branch of the Hekikai Shinkin bank. Photograph courtesy of Accsys

20. Amsterdam Marine Base

ARCHITECT:
Architects Bureau SLA

COMPLETION DATE:
2016

LOCATION:
Amsterdam (NL)

DESCRIPTION OF THE CASE-STUDY:

Architects Bureau SLA were tasked with presenting a design accommodating new uses within the building. Additionally, the Marine Base had to be renovated in time to host the Dutch Presidency of the European Union in the first half of 2016. Located in the heart of the city, the design consists of a new layout, new services and new facades. The facades consist of large 3.5 x 3.5 meter triple glazed windows set into deep window bays. Due to its structural integrity and sustainable properties, Accoya® was used for the main windows. The pattern of the wood appears random at first glance, however, they represent an interpretation of all the European communities national flags.

REFERENCES:

https://www.accoya.com/projects/project/accoya-selected-for-the-marine-base-amsterdam-building/

Fig. 4.20 Amsterdam Marine Base. Photograph courtesy of Milad Pallesh

21. Unique Loadbearing Strawbale Dome

ARCHITECT:
Gernot Minke
and Bjorn Kierulf (Createrra)

COMPLETION DATE:
2010

LOCATION:
Hrubý Šúr (SK)

DESCRIPTION OF THE CASE-STUDY:

Unique load bearing straw bale dome is nowadays used by the architectural studio Createrra. It is built as a combination of wooden supporting structure, load bearing straw bale vaults and dome. Heat recovery ventilation, air distribution through wooden ring beam. Structure is covered with clay plaster, EPDM foil and green roof. Different clay plaster finishes and earth floor are used in the interior.

REFERENCES:

http://minke-strawbaledome.blogspot.com/2010/08/gernot-minkes-strawbale-dome.html

Fig. 4.21 Unique loadbearing strawbale dome. Photograph courtesy of Createrra

22. MAI Modulo Abitativo (Housig Module)

ARCHITECT:
DUOPUU,
Cnr-Ivalsa Wood Building Design Lab

COMPLETION DATE:
2010

LOCATION:
San Michele all'Adige (IT)

DESCRIPTION OF THE CASE-STUDY:

MAI Modulo Abitativo Ivalsa prototype was designed by DUOPUU and wood building design lab of CNR-IVALSA in collaboration with Ceii Trentino and support of Provincia Autonoma di Trento. The goal of the project were to design and build a small wooden prototype house that can be built and prefabricated, easily transported by truck to the final destination and finally assembled to form a real house in a few hours. All the construction phases in the production area have been optimised to reduce the waste of materials, control air and water pollution, and achieve a highly efficient building process with respect to the environment and the workers.

REFERENCES:

http://www.duopuu.eu/essays/2010/09/mai-modulo-abitativo-ivalsa/

Fig. 4.22 MAI housing module. Photograph courtesy of Romano Magrone

23. Duna - Bird Watching Tower

ARCHITECT:
Bergen School of Architecture,
Faculty of Architecture STU Bratislava,
Veronika Kotradyová

COMPLETION DATE:
2018

LOCATION:
Hrušovská zdrž, Kalinkovo (SK)

DESCRIPTION OF THE CASE-STUDY:
This student´s hands-on project with environmental added value is promoting nature watching. The structure consists of 95% wood. The beams were CNC cut and glued before being transported to the site. The dressing is made up of three crossed layers of spruce and local pine finished by pigmented oil paint. The project aims to motivate people to come and learn about the birds and nature.

REFERENCES:

https://www.archdaily.com/868106/duna-bergen-school-of-architecture-plus-slovak-university-of-technology-in-bratislava

https://archello.com/project/duna

https://arkitektur-n.no/prosjekter/duna-bird-watching#

Fig. 4.23 Duna—bird watching tower. Photograph courtesy of Veronika Kotradyova

24. Observation Tree House

ARCHITECT:
Atelje Ostan Pavlin

COMPLETION DATE:
2015

LOCATION:
Celje, Bohinj (SI)

DESCRIPTION OF THE CASE-STUDY:
The Tree Observatory is a public building, which serves as a platform for pedagogical and cultural activities, exhibitions and contemplation. The structure of the house is built between six trees. The geometry of the building has six arms and intermediate terraces is connected to it. The central space is illuminated with zenith light through the "dome". The house is made of coniferous wood cut in the city forest and its equipment is modular and mobile.

REFERENCES:

http://www.piranesi.eu/atelje-ostan-pavlin-observation-tree-house-celje-city-forest-and-bicycle-trail-bohinj-slovenia/

https://www.celje.si/en/card/city-forest-tree-house

Fig. 4.24 Observation tree house. Image courtesy of Aleksander Ostan (outdoor image) and Jure Kravanja (indoor images)

References

Aksamija A (2016) Integrating innovation in architecture. In: Design, methods and technology for progressive practice and research. Wiley, Chichester

Jadhav NY (2016) Green and smart buildings. In: Advanced technology options. Springer Nature Singapore

Chapter 5
Service Life Performance

Prediction is very difficult, especially if it's about the future.
Niels Bohr

Abstract A special focus of this chapter is directed into assessing the performance of façades along the service life of the building. Influence of biotic and abiotic factors and their effect on materials physical and aesthetical properties are discussed. Principles of protection by design and their role in building performance during use phase are briefly introduced. Various approaches for the prediction of service life performance are supported with real case study data.

Façades directly influence building safety, comfort, and aesthetics. According to Martinez et al. (2015), façades have a role in security, thermal and acoustical insulation, air and water infiltration/mitigation, solar, daylight and glare control as well as in aesthetics. Over time, façade complexity evolved to accommodate a wide range of functionalities. Consequently, proper façade design has become a challenging and demanding task. Flores-Colen and de Brito (2010) listed several difficulties related to the performance of building façades. Those drawbacks are mainly related to poor design of construction details, improper choice or/and application of materials, and inadequate maintenance. To minimize mistakes in façade construction, modern design standards follow current building codes and specifications. However, the choice of proper materials is still challenging. Designers are often not willing to implement novel solutions with improved durability and low maintenance requirements, because they are not confident in their performance. At the same time, regular maintenance of façades is undesired among building owners. Fortunately, biomaterials are continuously improving and becoming more resistant to deterioration. Recent improvements in quality of biomaterials enable contractors to guarantee longer periods without maintenance, as well as to schedule the maintenance in advance. It is commonly accepted that façades require profound maintenance and/or partial renovation/replacement after 20 to 30 years following the construction (Martinez et al. 2015).

By predicting service life, it is possible to calculate lifelong construction costs (Gobakken and Lebow 2010). The deterioration intensity of façades depends on

A. Sandak et al., *Bio-based Building Skin*, Environmental Footprints and Eco-design of Products and Processes, https://doi.org/10.1007/978-981-13-3747-5_5

several variables, including building location, orientation, architectonical details, exposure level, microclimate, and intrinsic properties of materials used in construction. Clearly, all building materials (including concrete, metal, stone, and glass) require periodic cleaning, proper maintenance, and, finally, replacement. However, knowing which materials constitute façades is not enough to reliably predict service life performance of buildings, despite the great effort invested in trying (Fagerlund 1985).

5.1 Service Life Definition

The reference service life (RSL) is defined as a "service life of a product, component, assembly or system that is known to be expected under a particular set (e.g., reference set) of in-use conditions and which may form the basis of estimating the service life under other in-use conditions" (ISO 15686-1 2011). Evidently, service life of façades depends on several variables that are beyond the control of designers, such as local microclimate, environmental factors, or extreme weather events. However, regular inspection, maintenance, repair, or timely replacement of façades can significantly improve both building performance and owners' satisfaction. Proper balance between expected functionality, investment costs, and maintenance efforts enables sustained building performance. To assure optimal building functionality, it is critical to determine durability and service life of various materials, components, installations, and structures (Hovde 2002).

5.1.1 Service Life Categories

Kelly (2007) proposed division of the building service life into three main categories:

- Physical (technical) service life
- Functional service life
- Economic service life.

Physical or technical service life is related to the degradation of building elements due to natural ageing and deterioration factors. Probability of building failure increases systematically with time. This is mainly due to the natural ageing processes and degradation agents. Physical service life ends when the probability of building element failure is larger than the predefined level of acceptable risk.

Functional service life is linked to the expectations and demands of building users. In many cases, people are motivated to replace materials in order to follow recent trends and not because of their degraded technical performance (Rametsteiner et al. 2007). According to the studies conducted by Martinez et al.

(2015), aesthetics is the main motivation for retrofitting façades (accounting for 74% of all responses), followed by energy performance (65%), and remediation (56%). Therefore, trends in material choice should not be neglected when designing the building (Ebbert and Knaack 2007).

Economic service life is defined as the time between the beginning of building utilization and its replacement. Instead of maintaining existing façades, it is often both preferred and more cost effective to introduce new architectural solutions. Accordingly, buildings became economically obsolete with the development of new structural solutions that are more economical, more durable, and requiring less maintenance (Silva et al. 2016).

5.2 Deterioration of Biomaterials

Bio-based materials might provide optimal performance when assuring certain exposure conditions along the service life. Under specific circumstances, biomaterials may deteriorate to a different extent and velocity. Causes of degradation are generally divided into biotic and abiotic factors. Even though degradation mechanisms between these factors differ, both may occur simultaneously and affect degradation kinetics and the extent of the damage.

5.2.1 Biotic Factors

Fungi, insects, moulds, algae, and bacteria are the principal organisms degrading bio-based façade materials. Table 5.1 presents main types of damage caused by biotic agents and boundary climatic conditions stimulating their activity (Brischke et al. 2006; Viitanen et al. 2010). Detailed mechanisms of deterioration processes induced by organisms are described in the following part of the chapter. Severe decay of biomaterials is almost always caused by incorrect design, faulty assembly, or improper maintenance. To avoid deterioration by biotic agents, it is crucial to ensure proper drainage and drying in buildings (Clausen 2010). When it is not possible to maintain dry conditions, it is highly recommended to use either durable wood species or modified wood. Systematic inspection and repairs or replacements of decayed elements should be carried out regularly. Only dry and properly preconditioned elements without signs of fungal and insect infestations should be used as replacements. Obviously, direct causes of the initial decay (such as leakage, water trap, and perforation) should be eliminated to prevent the same problem appearing in the future.

Table 5.1 Key biotic agents causing deterioration of bio-based building materials

Organism	Damage	Environment cond. range	
		RH (%)	Temperature (°C)
Bacteria	Material softening, unpleasant odour, health problems	>97	−5 to 60
Mould	Discolouration, unpleasant odour, health problems	>75	0 to 45
Algae	Discolouration	>95	0 to 40
Decay fungi	Degradation of chemical components depending on the fungi type, material strength loss, unpleasant odour, health problems	>95	0 to 45
Insects	Strength loss (severity depending on insect taxon)	>65	5 to 50

5.2.2 Abiotic Factors

Abiotic factors are all inorganic chemical and physical components affecting buildings during their service life. They create boundary conditions for biotic agents. The origins of weathering are known; however, the mechanisms of degradation are only partially understood (Cogulet et al. 2018). Some of the most profound factors influencing the service life of biomaterials in façades are described below.

Water

Water, stemming from various sources (e.g., precipitation, condensation, leakage), is one of the key factors affecting weathering of biomaterials. Water that comes in contact with unprotected biomaterial surfaces is promptly absorbed by the sub-surface bulk, followed by further moisture migration within the cell walls (Feist and Hon 1984). When the relative humidity is increased, water vapour is directly absorbed by the bulk until the equilibrium moisture content is reached. The process of the moisture uptake (up to the level of fibre saturation) leads to material swelling. Due to the moisture gradient between the surface and its inner part, the material swells and shrinks, creating internal tensions and stresses. It may lead to the checking and cracking of biomaterials as well as membranes, when surfaces are coated. Important parameter that merges both climatic conditions and material deterioration is time of wetness (TOW). It indicates the period during which the atmospheric conditions promote the formation of a moisture layer directly wetting the biomaterial surface. TOW highly depends on the façade design, where horizontal surfaces limit the rainwater run-off. In general, TOW decreases with the increase of the exposure angle. TOW depends also on the wood species and modification processes applied (Van Acker et al. 2014; Meyer et al. 2012). In biomaterial surfaces, rain and wind remove residuals of photodegradation and erosion caused by weathering.

Solar Radiation

Biomaterial surfaces exposed to sunlight obtain visible colour alternations. In softwood, surface colours become more yellow or brown at the initial phase of degradation, followed by greying of the surface layer, as illustrated in Fig. 5.1. The colour changes occur to a depth of 0.05–2.5 mm, depending on the species and its specific density (Feist and Hon 1984). Discolouration is related primarily to the lignin decomposition resulting from the photodegradation. Even though discolouration kinetics differ between early and late wood, colour is nearly the same in both after prolonged weathering periods. Different parts of sunlight spectrum—visible, infrared, and UV light—degrade polymers; however, UV is the most damaging among them. Sunlight and water are considered the main factors affecting weathering intensity and kinetics. Interestingly, sunlight can both intensify and alleviate damaging effects of water (Feist and Hon 1984).

Temperature, Its Amplitude, and Gradient

The temperature of building surfaces is influenced by both external climatic conditions and properties of materials used. External factors include surrounding air temperature, solar radiation cumulative energy, relative air humidity, as well as wind direction and speed. Additional factors are the building orientation and its surroundings, presence of other buildings nearby, vegetation, and the proximity of water reservoirs. Material properties that influence temperatures of building surfaces include specific heat capacity, surface emissivity, radiation absorptivity and transmittance, albedo, bulk density, colour, and roughness. Since these properties vary among materials, it is clear that material choice is important in thermal performance of buildings. Materials can either decrease or increase near-surface air temperature, depending on how the bulk surface interacts with the sunlight (i.e., reflection, absorption, transmission). Consequently, façade materials play a major role in controlling ambient microclimates by affecting the temperature on the inside and the outside of buildings as well as inside urban spaces (Al-hafiz et al. 2017).

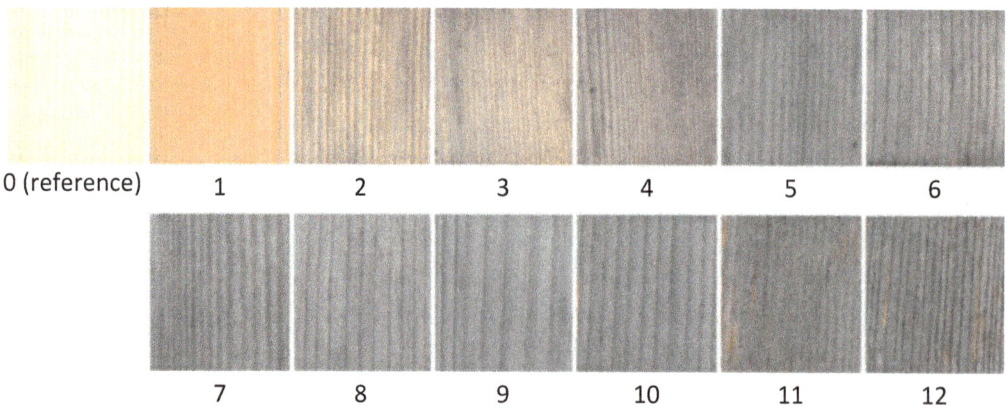

Fig. 5.1 Appearance of samples due to natural weathering (northern exposure) for 1 year in San Michele, Italy (numbers correspond to the months of exposure)

Therefore, in biomaterials that are considered for the use in façades, it is important to assess emissivity since it varies significantly among materials. A simple experimental approach to determine emissivity of biomaterials has been reported by Grossi et al. (2018).

Pollutants

Pollutants may be dispersed in the air as gas molecules, liquid microdrops, or solid particles. Even though most of the air pollution originates from man-made processes (e.g. industrial emissions, agriculture, transport), air is also polluted by natural processes, such as dust forming from rocks, volcanic activities, and soil erosion. According to Kuzmichev and Loboyko (2016), building façades are considerably degraded by dust deriving from polluted air. Especially, the effect of corrosive solid particles, liquid, and/or gaseous contaminants combined with a high humidity of ambient air has a highly negative impact on the external appearance of buildings (Kuzmichev and Loboyko 2016). For this reason, façades made of stone or brick should be regularly cleaned. Paints or coatings containing TiO_2 nanoparticles (or other hydrophobic agents) are recommended façades to sustainably and inexpensively reduce pollution and boost "self-cleaning" of façades (Mansour and Al-Dawery 2017). Moreover, according to recent studies, coating materials containing TiO_2 accelerate natural oxidation process of nitrogen oxides (NO_x) and sulphur oxides (SO_x) due to highly photocatalytic action of TiO_2 (Shukla et al. 2018). Hence, their use might help in reducing pollution and improving air quality in highly urbanized cities.

Physical Damage

Mechanical agents impose physical forces on buildings leading to stress and strain concentrations. The mechanical action can be static and permanent (such as ground pressure), or dynamic. In the second case, it is important to consider the influence of the wind and surface erosion caused by abrasion of suspended particles. The snow load is a static but temporary mechanical action also affecting the stress distribution within the structure. Even though limited mechanical deformations do not affect the structure integrity, they may stimulate generation of microfractures allowing water penetration and intrusion of micro-organisms. Façades, as an external part protecting the building, should be designed to resist such mechanical actions with minimal physical damage.

5.3 Deterioration Processes

By interacting with abovementioned biotic and abiotic factors, biomaterial surface and bulk undergo several changes affecting the original material properties. The mechanism of such changes is rather complex and may result from diverse processes occurring simultaneously or in sequence. The most pronounced processes of biomaterial degradation include ageing, weathering, mould and algae growth, decay, waterlogging, and insect infestation.

5.3.1 Ageing

Biomaterials start ageing immediately after their components, that is, plants, are cut down. The ageing process is highly influenced by environmental conditions (Fengel 1991). In timber, ageing is slow and occurs only in dry exposed wood structures, where the environmental conditions protect the material from fungi growth. Froidevaux and Navi (2013) defined ageing in wood as a slow chemical reaction taking place without the biotic degradation. During natural ageing, chemical reactions occur gradually, and the process may last for centuries. Changes in chemical composition of aged biomaterials depend on the species, environmental exposure, and the temperature–moisture history of the material. Fengel and Wegener (1989) noticed considerable decrements of the hemicellulose content in softwood roof elements aged for more than 300 years. The overall cellulose content in the wood does not change with ageing. However, due to the slow hydrolysis reaction, cellulose chains (microfibrils) were shortened and consequently polymerization degree decreases. Façades rarely age uniformly, since their exposure to environmental factors makes them prone to several degradation factors.

5.3.2 Weathering

Weathering is defined as a degradation process of materials that are exposed to external climatic conditions. In most cases, weathering is caused by abiotic factors and manifests as an alteration of colour, surface cracking, change in glossiness, and increased surface roughness. The intensity and extent of weathering depend on the local microclimate conditions, weather history, material type, function of an element, architectural details, and specific surface properties. The most important factors affecting weathering kinetics are solar radiation and stresses imposed by the cyclic wetting (moisture), together with temperature changes, environmental pollutants, and actions of certain micro-organisms. In general, unprotected raw biomaterials weather fast in the sub-surface, leaving the bulk intact. Differences in discolouration may be very noticeable within the same façade, especially in buildings where water and solar radiation are not spread uniformly across the surface (Fig. 5.2).

Fig. 5.2 Non-uniformly weathered building. Right image courtesy of Lone Ross Gobakken, NIBIO

Fig. 5.3 Weathered building in Stavanger, Norway. Image courtesy of Lone Ross Gobakken, NIBIO

Surface erosion in biomaterials is a side effect of weathering, where certain fibres resulting from the photodegradation of lignin are washed out. As late and early woods differ in their resistance to the erosion, surfaces can become curved, since low-density earlywood is removed faster. In mild climates, middle-density softwood generally erodes slowly, around 6 mm per century (Williams 2005). Often, weathering is a main degradation process affecting both the overall façade performance and the aesthetical perception of the building. However, in some cases weathering highlights building architectonic character and is not considered negative (Figs. 5.3 and 5.4).

Fig. 5.4 Astrup Fearnley Museum by Renzo Piano. Right image courtesy of Lone Ross Gobakken, NIBIO

5.3.3 *Moulds and Algae Growth*

Although mould does not affect the structural properties of wooden façades, it has a large effect on the aesthetical service life of the cladding. The main factors supporting its growth are water, nutrient source, and suitable temperature. The presence of moulds significantly changes façade appearance (Fig. 5.5). Colour of mould ranges from white, yellow, green, pink, brown to black. Some moulds (e.g., Trichoderma, Aspergillus) can leave a permanent stain or discolouration on the façade surface even after removed. Coatings may prevent mould growth very

Fig. 5.5 Wooden façade with the signs of moulds growth. Images courtesy of Lone Ross Gobakken, NIBIO

Fig. 5.6 Algae growth on the building façade

differently when exposed outdoors depending on the formulation and range of application (Gobakken et al. 2010). Since moulds and their spores are widely present in the environment, appropriate water management, adequate ventilation, and vapour barriers are necessary to protect building envelopes.

In addition to moulds, aerophilic algae frequently grow on building façades. Algae can develop in environments with high moisture levels (up to 100% RH), wide temperature range (0–40 °C), and limited light availability. Most algae can withstand extreme conditions, including severe temperatures and temporary drought. Algae most frequently grow on northern façades, where the direct sunlight illumination is minimal (Lengsfeld and Krus 2004). According to Gobakken and Vestøl (2012), growth of algae and lichen, cracking, flaking, and colour changes may also influence the aesthetical appearance, but usually to a lesser extent than mould and blue stain fungi. Figure 5.6 presents an example of a wooden cladding with algae coverage appearing unexpectedly on the northern façade after only 2 years of service life.

5.3.4 Decay

Fungi degrade wood by decomposing its constitutive polymers. The exact mechanism of this process depends on the specific enzymatic system of fungi. Brown-rot fungi use cellulase enzyme and degrade primarily polysaccharides (cellulose and hemicellulose), while they barely impact the lignin structure. The colour of wood becomes darker (light brown), and multiple cracks appear on the surface. The cracks may not only appear along the fibre direction but also across the grain, creating cubical pattern of discontinuities (Fig. 5.7a). The removal of cellulose results in a drastic decrease of certain mechanical properties of materials, such as modulus of elasticity (MOE) and modulus of rapture (MOR) (Singh 2000).

White-rot fungi are micro-organisms capable of decomposing both carbohydrates and lignin. Selective white-rot fungi degrade lignin in woody plant cell walls relatively to a higher extent than cellulose, while non-selective fungi may degrade all chemical wood components at a similar rate, leading to a uniform decay of the cell wall (Sandak et al. 2013). According to Winandy and Morrell (1993), these fungi do not significantly affect specific gravity, equilibrium moisture content, and bending properties. A visible sign of their presence is a spongy texture of wood without cross-cracks (Fig. 5.7b). Soft-rot fungi are typically active in wet environments, making materials severely low in strength (Singh 2000). The main enzyme of these fungi is cellulase, which easily degrades cellulose. In wood, this weakens the structure of the secondary wall while the middle lamella between cells remains unaffected.

White- and brown-rot fungi negatively impact aesthetical perception of façades and, in extreme cases, affect the structural integrity of buildings (Fig. 5.8). The maintenance of rotten façade parts is highly problematic. Often, the only reasonable strategy is to replace the infected elements.

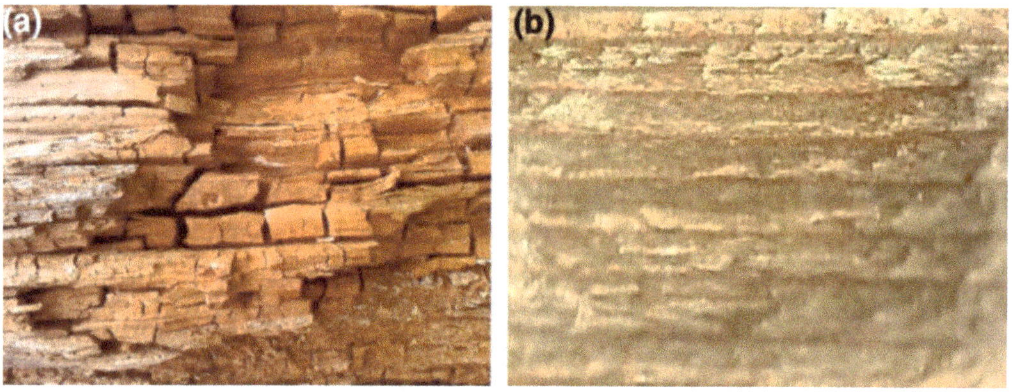

Fig. 5.7 Surface of biomaterial decayed by **a** brown-rot and **b** white-rot fungi

Fig. 5.8 Decay presence on the building façades. Images courtesy of Lone Ross Gobakken, NIBIO

5.3.5 Waterlogging

Waterlogging is typically irrelevant in building façades, unless the structures are erected in places directly affected by the water (including peats and wetlands) or exposed to frequent extreme weather events, such as floods. Most biomaterials slowly decompose in waterlogging conditions as a result of a complex interaction between the damp environment and the chemical constituents of the material. The key factors affecting the process are the presence (or absence) of oxygen, bacteria, fungi, and water temperature and its pH (Babinski et al. 2014; Sandak et al. 2014).

In general, chemical degradations slow down in the absence of oxygen, which can preserve biomaterials from extensive decomposition for millennia. However, certain bacteria are active also in anoxic environments. In this case, bacterial wood degradation is mainly stimulated by the water movement (Klaassen 2008). The most vulnerable part of the structure for degradation is the section that is at times immersed in water and at times exposed to ambient air (Fig. 5.9). In this case, the waterlogging degradation process caused by bacteria is far slower than the deterioration generated by fungi. This is because optimal conditions for fungal growth are just above the water level. Structures in such areas are delicate, and proper selection of construction materials is thus indispensable. It is recommended to use naturally durable wood species, treated/impregnated wood, or frequently replace façade elements at the water–air boundary. It is also important to design a structure in a way that the replacement of elements is simple.

Fig. 5.9 Wooden pier exposed to fluctuations of the water level. Image courtesy of Lone Ross Gobakken, NIBIO

Fig. 5.10 Wooden façade degraded by insects. Images courtesy of Magdalena Kutnik, FCBA

5.3.6 *Insects*

Most bio-based materials are food, shelter, and breeding environment for insects. Wood-destroying insects lay eggs in wood cracks and generally use wood as their main habitat. In wood, some of the most common insects are beetles (Coleoptera) and termites (Isoptera). Both belong to xylophagus, which means their diet is based on ligno-cellulosic materials (Brischke and Unger 2017). The larvae of beetles consume wood, producing a network of galleries which affect material mechanical properties and ease the access of other decaying micro-organisms. Termites living in large colonies (from 1000 to 2 million) are classified according to their living and feeding habits as dampwood, subterranean, and dry wood termites. Around 30 termite species are identified as causing damage to wooden buildings. It is not known in detail how different termite colonies choose their food. It is possible, however, to identify geographical locations where the termite attack is most probable. The risk of the insect infestation can be substantially reduced by implementing pest-resistant materials or impregnated/modified biomaterials identified as not attractive to xylophagus. An example of a wooden façade damaged by Cerambycidae, Anobiidae, and termites in presented in Fig. 5.10.

Marine borers (including mollusc or crustacean) may also be considered as organisms degrading biomaterials. However, their impact on building façades is minimal, except in waterlogging. In any case, using naturally resistant or impregnated biomaterials should be considered a necessity.

5.4 Potential Hazards and Degrading Agents

Other factors than biotic and abiotic should also be carefully considered when designing, installing, or using biomaterial façades. The most important factors are either anthropogenic or related to natural disasters.

5.4.1 Fire

When exposed to high temperatures, biomaterial properties change due to irreversible chemical reactions and physical changes of their chemical components. The extent of these changes depends on the temperature level and the exposure duration. At temperatures higher than 65 °C, strength and modulus of elasticity can permanently decrease, making the biomaterial more brittle. However, when exposed to high temperatures, biomaterials become a fuel and undertake the pyrolysis reaction when ignited. Following this, biomaterials develop an insulating layer of char that retards further degradation of sufficiently dense biomaterials, such as wood. The temperature in the char layer 6 mm below the burning surface is less than 180 °C due to the low thermal conductivity of wood. Therefore, bio-based structures can continue to carry a load even in the event of fire, assuming that their geometrical dimensions were sufficiently upscaled to assure the presence of uncharred cross sections (Dietenberger and Hasburgh 2016). In façades, knowledgeable design, implementation of proper materials, and careful installation may greatly influence the building performance during a fire. Bio-based materials in building façades can withstand a fire, if the façade geometry design does not permit contact between the fire plume and the cladding. Still, it is recommended to use materials treated with fire retardants and deflector elements able to change flame trajectory (Nguyen et al. 2016).

5.4.2 Flood

Bio-based materials fully immersed in water for prolonged periods will not decay, since the oxygen needed to sustain life of most micro-organisms is in shortage. On the other hand, biomaterials are more prone to biological attack immediately after the extensive wetting or flooding. The elevated risk lasts until the environmental conditions (i.e., air temperature, relative humidity, and biomaterial moisture content) return to non-hazardous levels. It should be considered that biomaterials tend to swell when saturated with water and may interact with other building components through an enormous pressure generated by the swelling. When flooded, bio-based building façade should be cleaned of mud and soil and, if at all possible, quickly dried. The drying process will shrink and further distort materials/structures. The extent of swelling and shrinkage depends on the species of which a biomaterial is derived from, biomaterial type, and the duration of water exposure (MacKenzie 2015).

5.4.3 *Earthquake*

Earthquakes generate ground motions in three dimensions that vary with time. Seismic movements are classified according to the movement direction. Vertical ground motions (caused by the P waves) generate forces that either add to or subtract from the gravitational force. They cause a fractional decrease of material volume by compressing it. In this case, load-carrying building structures usually help to prevent the damage and façades do not significantly contribute to the building integrity.

The impact of façades, however, may be important in horizontal movement (caused mainly by the S waves) that generates the most damage. Horizontal movements have a relatively large amplitude and cause shear deformations. Therefore, the building strength depends on the strength of the envelope and its connections. A properly designed façade made of biomaterials may help in reducing shear deformations and absorbing the disseminated energy. As a rule of thumb, the base of walls needs to be securely attached to the building foundation, while the roof and the floor must be safely fastened to the walls (Graf and Seligson 2011). Unfortunately, destructive effects of earthquakes can be magnified by the subsequent dangerous events, such as spontaneous fires, floods, soil liquefaction, or avalanches.

5.4.4 *Vandalism*

In the context of building façades, vandalism is the intentional damage of someone's property without the owner's permission. Damage resulting from vandalism varies in type and extent. Often, it is not necessary to replace damaged elements, since it is possible to clean or repaint the façade surface. In fact, one of the advantages of biomaterial façades is the relatively easy removal of graffiti. In most cases, façade surfaces can be washed, sanded, or repainted. In addition, there are several products able to protect façades with anti-graffiti coatings applied directly on biomaterials.

5.5 Protection by Design

Protection by design is an approach in which the planning process includes careful analysis of both the specific properties of materials used and the tailored building detailing. The overall goal is to eliminate or minimize any negative influences of material natural drawbacks. In bio-based building façades, it is essential to assure a fast water release from the structure after the event of wetting (Lstiburek 2006). This can be achieved in two ways:

Fig. 5.11 Different strategies for rain deflection. Right image courtesy of Marta Petrillo, CNR-IVALSA

- By implementing large-scale building elements that deflect rain, such as large roof overhangs and eves, protruding slab elements, and cantilevers (Fig. 5.11)
- By careful detailing of the small-scale technical solutions, such as connections, overlaps, and water traps elimination (Fig. 5.12).

Bio-based façades should be designed to avoid water entrapment between elements and to enable unimpeded moisture run-off. For example, tongue and groove connections should be implemented with the grove always on the top or overlapped connections similar to those of roof tiles should be used (Fig. 5.13). In case the water is ever entrapped, the façade design should enforce the air movement to facilitate drying and water evaporation. Open crevices that might create thin slits should be avoided, as they may absorb the water. Desired water migration could be also driven by the pressure differences between the surface of the façade and its interior. In this case, the wind causes higher pressure on the exterior face of the wall. The mitigation action in order to avoid rain penetration might be pressure equalization (Kudder and Erdly 1998). An important part of the protection by design paradigm is the use of flashing over the opening (e.g., windows and doors). Proper flashing includes the layering of water-resistant membranes, where each successive layer moves the water further away from the structure (Magee 2012). An example of the building that was designed by considering the abovementioned rules is presented in Fig. 5.11 (right). Here, the roof is large and protects the façade from the direct rain and snow precipitation. The part of the façade close to the ground is made of stone that avoids rainwater uptake and protects building foundations.

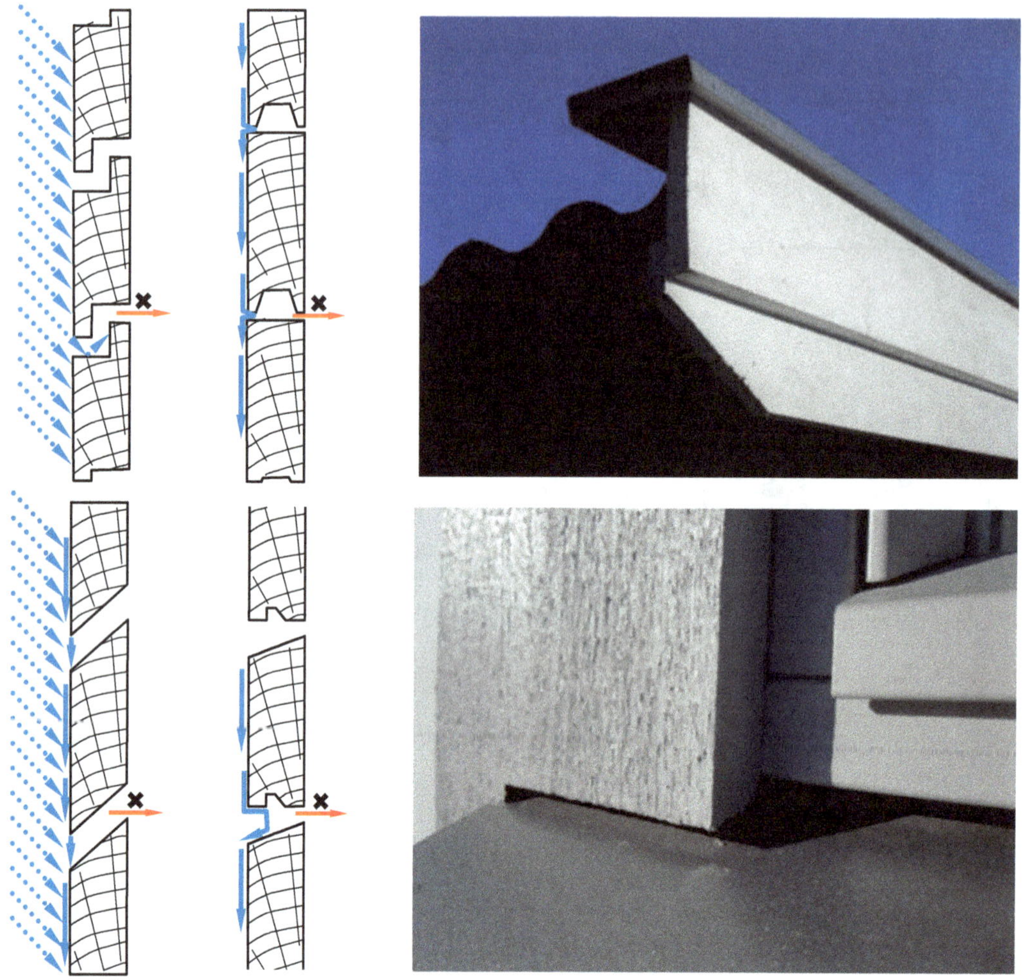

Fig. 5.12 Protection by design details allowing fast water run-off. Image courtesy of Lone Ross Gobakken, NIBIO

Fig. 5.13 Timber cladding with proper spacing between the timber planks and water run-off flashing

It is of utmost importance to foresee the maintenance schemes in different types of bio-based façades. All technical solutions in façades should enable easy dismantle of the modular façade units or at least allow simple replacement of single elements. Modular façades ease maintenance operations, such as sanding (removing the layer of corroded cellulose cells) or repainting/coating.

5.6 Serviceability, Durability, and Performance Over Time

To evaluate the overall building performance, the main criteria to consider are heat and mass transfer, acoustics, light access, fire resistance, costs, sustainability, and service life performance (Hendriks and Hens 2000). While the energy performance during the building use is often accounted for at the design phase simulations, the service life related to the biological, physical, and chemical attacks is often neglected. Service life of the building as a whole is influenced by several façade-related performance indicators, such as its durability (e.g., material quality, decay resistance, lifespan), compliance (e.g., maintenance flexibility, environmental impact), affordability (e.g., maintenance and refurbishment costs), and well-being (e.g., aesthetical appeal, customer satisfaction) (Jin et al. 2014). The estimated service life (ESL) prediction is a method useful in simulating total economic and environmental costs associated with the structure usage. The ESL of a structure ends when it reaches the predefined limit of acceptable imperfections or loses certain functions (Fig. 5.14). Ordinarily, the end of ESL is induced by aesthetical deterioration, long before the structure is deprived of its functions or safety. However, without the proper maintenance, the safety of a structure can be greatly diminished. The service life can be prolonged if the elements are replaced on time. In fact, service life can be substantially extended when maintenance is regularly scheduled and properly implemented. It is challenging, however, to maintain acceptable levels of appearance, functionality, and safety, while satisfying performance requirements and staying within reasonable limits of operational and maintenance costs.

5.7 Methods for Estimating Service Life Duration

Service life duration can be estimated with different methods. They produce different types of information, from absolute values to probabilistic distributions that characterize the estimated service life (Silva et al. 2016). Their advantages, limitations, and the quality of information produced are presented below.

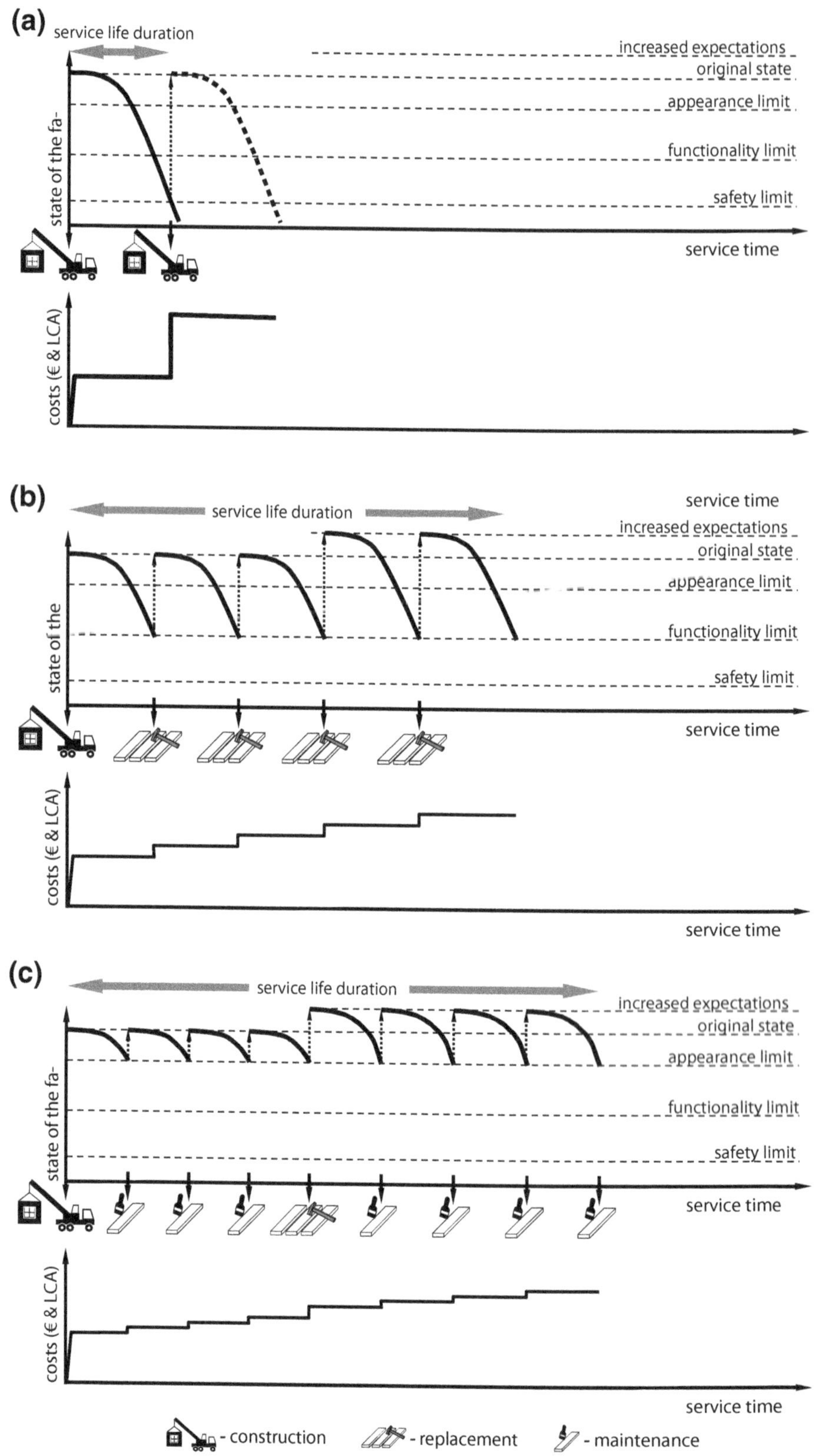

Fig. 5.14 Estimated service life of building façade according to different scenarios; no any repairs (**a**), replacement (**b**), and frequent maintenance (**c**)

5.7.1 Factorial Methods

According to the ISO standard (ISO 15686-1 2011), factorial method is the most frequently used method in recent times. It is based on calculating reference service life with consideration of several factors associated with specific usage conditions (Hovde 2002, 2005). This method considers service life characteristics of constructional elements but neglects degradation conditions. Unfortunately, small miscalculations are enough to significantly affect the service life estimation. Another drawback of the factor method is that it provides a single invariable value without considering possible fluctuations (Silva et al. 2016).

An adaptation of the ISO 15686 standard method used in exterior coated wood was extensively researched within the SERVOWOOD project (www.servowood.eu). There, the key deliverable was a new concept for the service life prediction (SLP) modelling approach using a customized set of "modifying" factors. Each of these factors reflects the effect of certain components on the performance of the wood coating system when compared to the reference coating solution. The weight of correction factors is determined experimentally by conducting long-term natural weathering tests. The SERVOWOOD method is expressed numerically in Eq. 5.1:

$$\mathrm{SLP} = \mathrm{RSL} \times A \times B \times C \times E \times F \times G \quad (5.1)$$

RSL reference value of service life length (years) estimated based on practical experience or experimental data;
A coating properties;
B substrate properties;
C production quality;
E exposure dose (outdoor environment);
F situation of component (usage conditions);
G inspection/maintenance measures.

5.7.2 Statistical Methods

Deterministic methods use mathematical and/or statistical formulations to describe the relationship between the degradation factors and the building condition. These methods intend to obtain a function that best fits a set of random data. For this, they require sets of large and representative data to be used in numerical/statistical modelling.

Regression can be used to model weathering deterioration progress and, consequently, service life prediction. Here, partial least squared (PLS) regression or multi-way data analysis can be carried out. In example proposed by Sandak et al. (2015a, b), PLS evaluates characteristics of two extreme samples, one not at all exposed to weathering and another that is at the final stage of the weathering deterioration. A non-exposed sample is assigned the degradation index 0, while a completely degraded samples receives the index 1000. PLS incorporates reference data corresponding to both extremes of deterioration levels. The data include objectively assessed characteristics, such as colour, gloss, chemical composition, and roughness. PLS can also incorporate new sets of specific characteristics and thus estimate the progress of weathering in previously unknown samples (Sandak et al. 2015a; b, Sandak et al. 2016). This method can predict the degradation level by placing a sample somewhere in between the brand new and completely degraded. However, the model requires all samples to follow the same degradation mechanism/path and separate PLS models have to be developed for different sample configurations (e.g., material type, coating, exposure conditions).

As an alternative to PLS, multi-way data analysis (MWA) is a method of analysing information sets that can be represented in a multi-dimensional array (Sandak et al. 2017). The dimension may correspond to either measurement duration, location, or methodology/parameter. MWA methods are particularly useful in time series analysis and, consequently, in deterioration/weathering modelling. Variables (sample characteristics) acquired during the weathering in each experimental location are portrayed in a multi-way array. A result of the MWA is a set of loadings and scores for each mode, case, and variable. Loadings can be examined separately to enhance the understanding of the influences of experimental variables (e.g., exposure location, test duration, material properties, surface finishing). Like PLS, the MWA model can predict the weathering of unknown samples. It can also simulate the weathering progress in numerous weathering scenarios.

5.7.3 Experimental Methods

Experimental methods are the alternative approaches to those presented above. These methods typically carry out in situ measurements or accelerated tests in laboratory conditions in order to evaluate effects of specific agents on the deterioration. Empirical methods are usually augmented by in-field survey of degradation factors leading to the determination of the life cycle expectancy (Flores-Colen and de Brito 2010). However, acquiring data in real service life conditions is challenging, since it is often not practical to expose samples to deteriorating conditions for a sufficiently long duration (Galbusera et al. 2014).

Unfortunately, the abovementioned modelling methods do not directly link degradation of samples exposed to natural weathering to the local climate history. Diverse weather factors may cause numerous alterations to the weathered material. Dose–response relationships should be used to explain the biomaterial façade degradation in relation to varying exposure conditions (Petrillo et al. 2019). A procedure in developing a dose–response model typically includes the following steps. First, relevant meteorological data have to be collected and homogenized (Sandak et al. 2015b). These data are used to compute the weather dose separately for each location and time period. Second, the degradation index is computed to characterize the degradation progress at different stages of the weathering test. This process is based on the characteristics of weathered materials measured during the weathering test. The degradation index represents material response to conditions causing deterioration. Both weather doses and corresponding responses are used by the iterative modelling algorithm to develop the numerical model. The model can be used to simulate the façade service life and to determine when in the future will the building lose its aesthetical appeal for future occupants (Sandak et al. 2015b). This information can be crucial in the maintenance planning and in the assessment of environmental and economic costs of the structure during its service life.

Performance of façade materials depends on various factors, whose effects on deterioration, physical–chemical mechanisms, inhibitions, and cross-correlations are often uncertain (Verma et al. 2014). Models based on the probability theory or statistical concepts are suitable for describing such complex processes. Typical steps in such methods are (1) defining failure limit states for the corresponding features, (2) quantifying random variables, and (3) calculating probability levels (Li and Vrouwenvelder 2007).

5.7.4 Stochastic Methods

Stochastic methods provide information related to the risk of failure, where the most probable failure time is predicted by taking into account selected building elements (Silva et al. 2016). The most common examples of stochastic methods are logistic regressions and Markov chains. Logistic regression evaluates the probability of transition between degradation conditions over time, considering specific façades characteristics. It estimates the time when the façade will reach the end of its service life. The Markov chain provides similar results but analyses processes selectively. Markov chain cannot produce context-dependent results, since it does not consider the overall context and the entire range of circumstances. The phases of the deterioration (degradation) process are analysed separately and not sequentially. Stochastic methods provide valuable information, especially for identifying maintenance strategies that should be implemented in façades during their service life.

5.7.5 Computational Methods

Computational methods rely on mathematical models to describe complex systems with computer simulations. These methods use existing data to find the best fitting models. Such approach is particularly useful in predicting the behaviour of building elements, especially when intuitive solutions are not available. Computational methods perform well when dealing with inaccurate data and samples that contain outliers. The most popular techniques are artificial neural networks (ANNs) and fuzzy systems (fuzzy logic). ANN uses existing empirical knowledge to model the reality. Specifically, by applying specific learning algorithms, it transforms raw data into models approximating real-life phenomena. Compared to conventional linear models, fuzzy logic models are better equipped to deal with the uncertainty associated with complex phenomena, such as degradation of construction elements (Silva et al. 2016).

5.7.6 Visualization and Simulation Methods

Aesthetics is highly subjective. To estimate aesthetic service life, specific requirements and accurate prediction of the visual appearance change should be considered (Lie et al. 2018). Numerical modelling can be used not only to anticipate the expected service life but also to schedule the maintenance activities in order to preserve façade aesthetical qualities. The model can simulate changes in façade appearance caused by surface weathering, which is the main process affecting the deterioration of façades (Sandak et al. 2017). The time to the first repair can be estimated in various scenarios while considering a wide array of variables, such as different aesthetical preferences of building owners, diverse architectural solutions, or specific microclimate conditions. When this tool is integrated with the modern building information modelling software, its scope broadens even further. For instance, it can simulate façade appearance (during the entire service life) taking into account the choice of biomaterials. It is an excellent platform to determine aesthetical limit states and environmental impact of façades, both during and after the use phase. It is also an indispensable tool in predicting investment and maintenance costs, considering various scenarios of maintenance and/or replacement. Such modelling tools were developed in BIO4ever project (www.bio4everproject.com) and are currently undergoing extensive testing and validation (Sandak et al. 2018).

5.7.7 ESL Models Update and Validation

The capacity and reliability of a model to represent real-life systems must be confirmed in validation procedures. Ideally, a model should be able not only to describe the variability of the data from the teaching data set, but it should also be applicable to other scenarios. To achieve this, it is important to continuously upgrade numerical models by implementing additional data gradually acquired from newly erected structures. Furthermore, the database of façade materials should be regularly updated, especially by including new bio-based building materials emerging on the market (Brischke and Thelandersson 2014). The most efficient software systems can automatically acquire new data during building service life and incorporate it into the original model. There are different model validation strategies, depending on the extent of the available cases and on the future accessibility to new independent data. However, external validation with independent data is always considered an optimal solution and thus recommended for routine implementation.

5.7.8 Concluding Remarks

Existing methods in predicting ESL and modelling long-term degradation and performance of building façades vary in their complexity and reliability. Some of these (e.g., factorial method) tend to oversimplify degradation processes, since they do not always consider (1) the full range of variability of factors involved in the ageing of the structure and (2) the specificity of the modelled component. Other methods, such as stochastic, are too complex to be routinely used, since they describe the service life in too specific conditions and are mostly based on the probability theory (Moser and Edvardsen 2002). Computer-based simulation tools require exceptional skills in using certain software and in building physics. For this reason, most of the available tools are developed and used mainly for research purposes (Vullo et al. 2018). Thus, it is essential to develop simple but reliable modelling tools that integrate all the abovementioned methods. For more details and examples of modelling to predict expected service life, readers are encouraged to refer to the work of Masters (1985), Saunders (2007), Silva et al. (2016), and Jones and Brischke (2017).

References

Al-hafiz B, Musy M, Hasan T (2017) A study on the impact of changes in the materials reflection coefficient for achieving sustainable urban design. Procedia Env Sci 38:562–570

Babiński L, Izdebska-Mucha D, Waliszewska B (2014) Evaluation of the state of preservation of waterlogged archaeological wood based on its physical properties: basic density vs. wood substance density. J Archaeol Sci 46:372–383

Brischke C, Rapp AO, Bayerbach R (2006) Decay influencing factors: a basis for service life prediction of wood and wood-based products. Wood Mat Sci Eng 1:91–107

Brischke C, Thelandersson S (2014) Modelling the outdoor performance of wood products—a review on existing approaches. Constr Build Mater 66:384–397

Brischke C, Unger W (2017) Potential hazards and degrading agents. In: Jones D, Brischke C (eds) Performance of bio-based building materials. Elsevier, pp 188–203

Clausen CA (2010) In: Ross RJ (ed), Biodeterioration of wood in wood handbook—wood as an engineering material. Forest Products Laboratory, Madison

Cogulet A, Blanchet P, Landry V (2018) The multifactorial aspect of wood weathering: a review based on a holistic approach of wood degradation protected by clear coating. BioRes 13 (1):2116–2138

Dietenberger MA, Hasburgh LE (2016) Wood products thermal degradation and fire. In: Reference module in materials science and materials engineering, 8 pp

Ebbert T, Knaack U (2007) A flexible and upgradeable façade concept for refurbishment. In: Bragança L et al (ed) Proceedings of the SB07 Lisbon–sustainable construction, materials and practices: challenge of the industry for the New Millennium, Rotterdam

Fagerlund G (1985) Essential data for service life prediction. In: Problems in service life prediction of building and construction materials. NATO ASI series, 95, pp 113–138

Feist WC, Hon DNS (1984) Chemistry of weathering and protection. In: Rowell RM (ed) Chemistry of solid wood. Advances in Chemistry Series 207, A.C.S., Washington, pp 401–451

Fengel D (1991) Aging and fossilization of wood and its components. Wood Sci Technol 25: 153–177

Fengel D, Wegener G (1989) Wood: chemistry, ultrastructure, reactions. De Gruyter Publications, Berlin

Flores-Colen I, de Brito J (2010) A systematic approach for maintenance budgeting of buildings façades based on predictive and preventive strategies. Constr Build Mater 24(9):1718–1729

Froidevaux J, Navi P (2013) Aging law of spruce wood. Wood Mat Sci Eng 8(1):46–52

Galbusera MM, de Brito J, Silva A (2014) The importance of the quality of sampling in service life prediction. Constr Build Mater 66:19–29

Gobakken LR, Lebow PK (2010) Modelling mould growth on coated modified and unmodified wood substrates exposed outdoors. Wood Sci Technol 44:315–333. https://doi.org/10.1007/s00226-009-0283-0

Gobakken LR, Høibø OA, Solheim H (2010) Mould growth on paints with different surface structures when applied on wooden claddings exposed outdoors. Int Biodeterior Biodegradation 64:339–345

Gobakken LR, Vestøl GI (2012) Surface mould and blue stain fungi on coated Norway spruce cladding. Int Biodeterior Biodegradation 75:181–186

Graf WP, Seligson A (2011) Earthquake damage to wood-framed buildings in the shakeout scenario. Earthquake Spectra 27(2):351–373

Grossi P, Sandak J, Petrillo M et al (2018) Effect of wood modification and weathering progress on the radiation emissivity. In: 9th European Conference on Wood Modification, 17–18 September 2018, Arnhem, Netherlands

Hendriks L, Hens H (2000) Building envelopes in a holistic perspective. Laboratorium Bouwfysica, Amsterdam, Netherlands

Hovde PJ (2002) the factor method for service life prediction from theoretical evaluation to practical implementation. In: Proceedings of the 9DBMC international conference on durability of building materials and components, Brisbane, Australia 10p

Hovde PJ (2005) the factor method—a simple tool to service life estimation. In: 10 DBMC international conferences on durability of building materials and components, Lyon, France

ISO 15686-1 (2011) Buildings and constructed assets—service life planning—part 1: general principles and framework. International Organization for Standardization

Jin Q, Overend M (2014) A prototype whole-life value optimization tool for façade design. J Build Perform Simul 7(3):217–232

Jones D, Brischke C (eds) (2017) Performance of bio-based building materials. Elsevier

Kelly DJ (2007) Design life of buildings—a scoping study. Scottish Building Standards Agency

Klaassen RKWM (2008) Water flow through wooden foundation piles: a preliminary study. Int Biodeterior Biodegrad 61(1):61–68

Kudder RJ, Erdly JL (1998) Water leakage through building facades. ASTM International

Kuzmichev AA, Loboyko VF (2016) Impact of the polluted air on the appearance of buildings and architectural monuments in the area of town planning. Procedia Eng 150:2095–2101

Lengsfeld K, Krus M (2004) Microorganism on façades—reasons, consequences and measures. IEA ANNEX 41 meeting, Glasgow, Scotland

Li Y, Vrouwenvelder T (2007) Service life prediction and repair of concrete structures with spatial variability. Heron 52(4):251–267

Lie SK, Vestøl GI, Høibø O, Gobakken LR (2018) Surface mould growth on wooden claddings—effects of transient wetting, relative humidity, temperature and material properties. Wood Material Science & Engineering. https://doi.org/10.1080/17480272.2018.1424239

Lstiburek J (2006) Understanding basements. Building Science and Corporation

MacKenzie C (2015) Impact and assessment of moisture-affected timber-framed construction. Forest and Wood Products Australia Limited. ISBN 978-1-921763-42-7

Magee M (2012) Wood Protection by Design. Timber Framing 103:21–27

Mansour AMH, Al-Dawery SK (2017) Sustainable self-cleaning treatments for architectural façades in developing countries (in press)

Martinez A, Patterson M, Carlson A et al (2015) Fundamentals in façade retrofit practice. Procedia Eng 118:934–941

Masters LW (ed) (1985) Problems in service life prediction of building and construction materials. Springer, Dordrecht

Meyer L, Brischke C, Pilgård A (2012) Modified timber in various above ground exposures—durability and moisture performance. In: Hill CAS, Militz H, Pohleven F (eds) Proceedings of the 6th european conference on wood modification. Ljubljana, Slovenia

Moser K, Edvardsen C (2002) Engineering design methods for service life prediction. In: 9DBMC international conference on durability of building materials and components, Brisbane, Australia

Nguyen KTQ, Weerasinghe P, Mendis P et al (2016) Performance of modern building façades in fire: a comprehensive review. Electron J Struct Eng 16(1):69–86

Petrillo M, Sandak J, Grossi P et al (2019) Chemical and appearance changes of wood due to artificial weathering—dose-response model. J Near Infrared Spectrosc. https://doi.org/10.1177/0967033518825364

Rametsteiner E, Oberwimmer R, Gschwandtl I (2007) Europeans and wood: what do Europeans think about wood and its uses? A review of consumer and business surveys in Europe. ISBN: 978-83-926647-0-3

Sandak A, Ferrari S, Sandak J et al (2013) Monitoring of wood decay by near infrared spectroscopy. Adv Mater Res 778:802–809

Sandak A, Sandak J, Babinski L et al (2014) Spectral analysis of changes to pine and oak wood natural polymers after short-term waterlogging. Polym Degrad Stab 99:68–79

Sandak J, Sandak A, Riggio M (2015a) Multivariate analysis of multi-sensor data for assessment of timber structures: principles and applications. Constr Build Mater 101(2):1172–1180

Sandak J, Sandak A, Riggio M (2015b) Characterization and monitoring of surface weathering on exposed timber structures with multi-sensor approach. Int J Archit Heritage 9:674–688

Sandak A, Sandak J, Burud I et al (2016) Weathering kinetics of thin wood veneers assessed with NIR. J Near Infrared Spectrosc 24(6):549–553

Sandak J, Sandak A, Burud I (2017) Modelling of weathering. In: Jones D, Brischke C (eds) Performance of bio-based building materials. Elsevier, pp 502–510

Sandak J, Sandak A, Grossi P et al (2018) Simulation and visualization of aesthetic performance of bio-based building skin. In: Proceedings IRG annual meeting. ISSN 2000-8953, IRG/WP 18-20633

Saunders SC (2007) Reliability, life testing and the prediction of service lives. Springer, Dordrecht

Shukla A, Waris RF, Arhab M et al (2018) A review on pollution eating and self-cleaning properties of cementitious materials. Int J Innov Res Sci Eng Technol 7(3):2784–2792

Silva A, de Brito J, Gaspar PL (2016) Methodologies for service life prediction of buildings with a focus on façade claddings. Springer International Publishing, Switzerland

Singh J (2000) Fungal problems in historic buildings. J Archit Conserv 6:17–37

Van Acker J, De Windt I, Li W et al (2014) Time of wetness (ToW) simulation based on testing moisture dynamics of wood. In: COST FP1303: performance and maintenance of bio-based building materials influencing the life cycle and LCA, Kranjska Gora

Verma SK, Bhadauria SS, Akhtar S (2014) Probabilistic evaluation of service life for reinforced concrete. Struct Ch J Eng, Article ID 648438, 8 p

Viitanen H, Vinha J, Salminen K et al (2010) Moisture and bio-deterioration risk of building materials and structures. J Build Phys 33(3):201–224

Vullo P, Passera A, Lollini R et al (2018) Implementation of a multi-criteria and performance-based procurement procedure for energy retrofitting of façades during early design. Sustain Cities Soc 36:363–377

Williams RR (2005) Weathering of wood. In: Handbook of wood chemistry and wood composites edited by Rowell. R. M. CRC Press

Winandy JE, Morrell JJ (1993) Relationship between incipient decay, strength, and chemical composition of Douglas fir heartwood. Wood Fiber Sci 25(3):278–288

www.bio4everproject.com. Accessed on 08 Oct 2018

www.servowood.eu. Accessed on 08 Oct 2018

Chapter 6
Portfolio of Bio-Based Façade Materials

Each material has its specific characteristics which we must understand if we want to use it.

Ludwig Mies van der Rohe

Abstract This chapter presents a selection of biomaterials identified by industry and academia as superior for building façades. Time series of photographs demonstrating changes of material appearance during use phase are provided for each case. In addition, selected technical characteristic, durability, recyclability potential as well as costing estimates are provided for each biomaterial.

The trend for rapid deployment of novel/advanced material solutions at reduced costs through predictive design of materials and innovative production technologies is observed nowadays. Such materials are optimized for specified applications, assuring at the same time expected properties and functionality at elongated life, minimizing the environmental impact, and reducing risk of product failure. As a consequence, higher numbers of well-performing (also in severe environments) construction materials are available on the market. It is extremely important for the biomaterial production sector to follow this trend and to continuously improve its offer. Today's bio-based building materials, even if well characterized from the technical point of view, are often lacking reliable models describing their performance during service life. The other factor, often underestimated (but critical for the sustainable use of bio-based building materials), is related to the transformations of building materials after their service life. The development of really innovative and advanced bioproducts relies on the deep understanding of the material properties, structure, assembly, formulation, and its performance along the service life.

A. Sandak et al., *Bio-based Building Skin*, Environmental Footprints and Eco-design of Products and Processes, https://doi.org/10.1007/978-981-13-3747-5_6

6.1 BIO4ever Project

The BIO4ever project was dedicated to fulfilling gaps of lacking knowledge on some fundamental properties of novel bio-based building materials. The two driving objectives of the projects were:

- To promote use of biomaterials in modern construction by understanding/modelling its performance as function of time and weathering conditions
- To identify most sustainable treatments of biomaterial residues at the end of life, improving even more their environmental impact.

The overall goal was to assure sustainable development of the wood-related construction industry, taking into consideration environmental, energy, socio-economic, and cultural issues. This has been achieved by developing original, trustworthy tools demonstrating advantages of using bio-based materials when compared to other building resources.

6.1.1 BIO4ever Project Materials

Hundred twenty samples investigated within the BIO4ever project were provided by 30 industrial and academic partners from 17 countries. The experimental samples include different wood species from various provenances, thermally and chemically modified wood, composite panels, samples finished with silicone and silicate-based coatings, nanocoatings, innovative paints and waxes, impregnated wood, bamboo composites, reconstituted slate made with bioresin, and samples prepared according to traditional Japanese technique: Shou Sugi Ban. The experimental samples were classified into seven categories, according to the treatment applied: natural wood (or other bio-based materials), chemically modified, thermally modified, impregnated, coated and/or surface treated, composites, and hybrid modified. Hybrid modification was defined here as a combination of at least two different treatments (Table 6.1).

6.2 Weathering of Biomaterials

In order to properly use biomaterials, it is indispensable to thoroughly understand their properties and performance in service. In any case, the actual performance highly depends on the local conditions, specific microclimate, and architectural details of the building (among the others). There are several commonly accepted protocols developed for comparison of the façade material performance and resistance to weathering, including two ways of testing:

Table 6.1 Categories of bio-based materials suitable for building façades tested within BIO4ever project

Category	Examples	Number of cases
Natural	Wood, bamboo	19
Impregnated	Furfurylation, DMDHEU, Knittex, Madurit, Fixapret	28
Thermally modified	Vacuum, saturated steam, oil heat treatment	20
Chemically modified	Acetylation	5
Coating and surface treatments	Different coatings, carbonized wood, nanocoatings	16
Hybrid modification	Thermal treatment + coating, thermal treatment + impregnation, acetylation + coating, etc.	25
Composites	Panels, bioceramics, Tricoya®, wood–plastic composites	7

- Natural weathering in the exterior
- Accelerated ageing performed usually in UV chambers with water spraying.

More details regarding technical requirements, interpretation of results, and comparison between diverse materials were described in detail in several publications, including Reinprecht (2016) and Jones and Brischke (2017).

6.2.1 *Natural Weathering*

Natural weathering procedure is defined in standard EN 927-3—Paints and varnishes—Coating materials and coating systems for exterior wood—Part 3: Natural weathering test (CEN EN 927-3 2014). This standard evaluates the resistance to natural weathering of the tested coating system, applied to a wood substrate. The durability is evaluated by determining the changes in decorative and protective properties of coatings after 12 months of exposure. In this case, samples are exposed on the racks, inclined at an angle of 45° to the horizontal level and facing south.

The list of surface aspects altered by the exposure to the weather conditions as recommended by the standard to evaluate along the test includes:

- Specular gloss to be assessed by gloss meter
- Colour and colour difference in CIE *Lab* colour coordinates measured by colorimeter or spectrophotometer
- Paint adhesion test by the pull-out or scratch trial
- Visual surface defect assessment with microscope of ×10 magnification
- Change of film thickness
- Chalking by self-adhesive, transparent tape.

Fig. 6.1 Natural weathering experiment conducted within BIO4ever project

All the biomaterials presented here undergo a natural weathering test conducted in San Michele all'Adige, Italy (46° 11′ 15″N, 11° 08′ 00″E) in the period from April 2016 to July 2018. An image of the experimental stands is shown in Fig. 6.1.

6.2.2 Artificial Weathering

Artificial weathering procedure is described in standard 927-6 "Paints and varnishes—Coating materials and coating systems for exterior wood—Part 6: Exposure of wood coatings to artificial weathering using fluorescent UV lamps and water" (CEN EN 927-3 2006). The standard defines the method for determining the resistance of wood coatings to artificial weathering performed in an apparatus equipped with fluorescent UV lamps, combined with system of vapour condensation and water spray. The laboratory test is carried out taking into consideration a limited number of variables (radiation, temperature, and humidity) which can be fully controlled and monitored. Accordingly, the artificial weathering test results are much more reproducible and comparable than these from natural weathering. The usual setting of the artificial weathering cycle contains 2.5 h of UV light irradiation (e.g., lamp UVA-340 simulating the sunlight radiation from 365 to 295 nm, with a peak emission at 340 nm) followed by 0.5 h of water spray. The standard unit test lasts for 168 h (1 week). However, it is recommended that such a test has to be continued for 12 times resulting in the total test duration of 2016 h (12 weeks). Figure 6.2 presents sample after 672 h of artificial weathering.

6.3 Characterization of the Biomaterial's Surface Properties

Investigated biomaterials should be characterized before, during, and after degradation by biotic and abiotic agents, in order to quantify the deterioration progress as well as to provide experimental data for further analysis. Such data are critical for

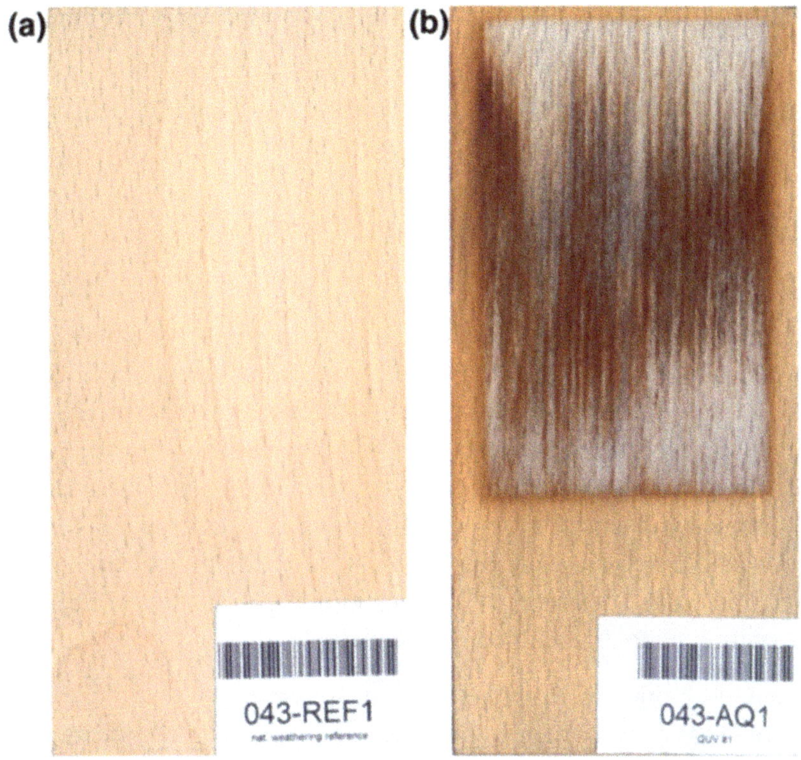

Fig. 6.2 Appearance of samples before and after artificial weathering

understanding the physical–chemical mechanisms of biomaterial degradation as a function of time and to model their performance during service life. The set of surface properties assessed for the needs of this portfolio included image acquisition, colour measurement, glossiness, and durability. The methodology and measurement protocols are described below.

6.3.1 Digital Colour Image

Digital colour images of samples surfaces were collected by means of HP G2710 office scanner. The sample was conditioned before scanning in the room conditions (temperature of 20 °C and 55% of relative humidity) in order to eliminate any effect of the moisture on the surface appearance. All the images were acquired with a resolution of 300 DPI.

6.3.2 Colourimetry

Changes in colour can be assessed by a spectrometer following the CIE *Lab* system where colour is expressed with three parameters:

- *CIE L**—correlated to lightness
- *CIE a**—defining red-green tone
- *CIE b**—defining yellow-blue tone.

Even if CIE *Lab* is not the only way to define colour, it is considered as an industrial standard when controlling quality or characterizing colour of surface. However, it is most suitable for measurement of homogenous surfaces and defines an average colour over the surface measured by the aperture of the entrance slit. It is impossible therefore to determine both, surface pattern and variability of colour within the sample by means of a single CIE *Lab* value. It is especially important in a case of wood and other biomaterial possessing very complex surface and pattern distribution. The colour determination procedure implemented here for assessment of the biomaterial samples included:

- Instrument calibration with white and dark references
- Setting of the measurement conditions and colour computation variables (Illuminant D65 and viewer angle 10°)
- Positioning the probe over the measured surface
- Measurement and data collection.

CIE *Lab* colours were measured using MicroFlash 200D spectrophotometer (DataColor Int, Lawrenceville, USA). All specimens were measured on ten different spots. The standard deviation of such measurement was considered as an indicator of the texture colour variability of the material.

6.3.3 Gloss

The light-irradiating surface is partially reflected following two physical principles: specular and diffuse reflections. The specular mode with incidence/reflectance angle of 60° was measured here by using REFO 60 (Dr. Lange, Düsseldorf, Germany) gloss meter. Ten measurements were taken on each specimen, following along the fibre direction.

6.4 Biotic Durability

The natural durability of a wood species is defined as its inherent resistance to wood-destroying agents. Two of the fundamental standards EN 350-1 (CEN 1994a) and EN 350-2 (CEN 1994b), developed by the CEN/TC 38, were recently replaced by revised standard EN 350 (CEN 2016) "Durability of wood and wood-based products—Testing and classification of the durability to biological agents of wood and wood-based materials" published in 2016. The standard covers wood-decaying fungi (basidiomycete and soft-rot fungi), beetles capable of attacking dry wood and termites and marine organisms capable of attacking wood in service. EN 350 provides the durability classes of wood-based materials to various wood-destroying organisms as summarized in Tables 2.1 and 2.2 (Chap. 2). Unfortunately, the standard EN 350 is not meant to test ligno-cellulosic materials other than wood. For that reason, custom methodological adaptations for testing other organic materials such as bamboo, reed, straw, or flax are proposed by diverse laboratories (Kutnik et al. 2014, 2017).

Field performance can be assessed by tests simulating in-service conditions. For example, EN 252 (CEN 2014) evaluates the performance of wood and wood-based materials to withstand biodeterioration. Evaluated materials in the form of stakes are half-buried in the soil and visually assessed and rated according to the attack severity. Additionally, mass loss is calculated according to Eq. 6.1:

$$\mathrm{ML} = \frac{m_0 - m_t}{m_0} \cdot 100\% \tag{6.1}$$

where ML is mass loss (%), m_0 is dry weight of the sample before the test (g), and m_t is dry weight of the sample after the test (g). Examples of durability test performed in the field and in the laboratory conditions according to EN 350 standard are presented in Fig. 6.3.

6.5 Portfolio of Selected Biomaterials Tested Within Frame of BIO4ever Project

The progress of natural weathering and appearance change of investigated samples are presented on the left part of Figs. 6.4, 6.5, 6.6, 6.7, 6.8, 6.9, 6.10, 6.11, 6.12, 6.13, 6.14, 6.15, 6.16, 6.17, 6.18, 6.19, 6.20, 6.21, 6.22, 6.23, 6.24, 6.25, 6.26, 6.27, 6.28, 6.29 and 6.30. Some materials appear to not change the outlook; however, several of these changed (usually become pale and less saturated), like in the case of natural wood or acetylated wood. Changes in material appearance (numerical data regarding colour and gloss) are summarized in the table situated on right, down side of each figure. Technical details regarding selected samples representing seven treatment categories are presented on the right upper side. Short materials' description is followed by a selection of the basic information regarding product availability on the market, estimated cost range, performance characteristics (durability against fungi, termites, wood-boring insects, and weather), fire resistance, recyclability potential, and maintenance effort. The meaning of symbols is described in Tables 6.2 and 6.3.

Fig. 6.3 Experimental samples under durability assessment: **a** tested in the laboratory against termites, **b** field test according to EN 252 (adapted to mini-size steaks), **c** samples after 3 months of laboratory test, **d** assessment of samples in field

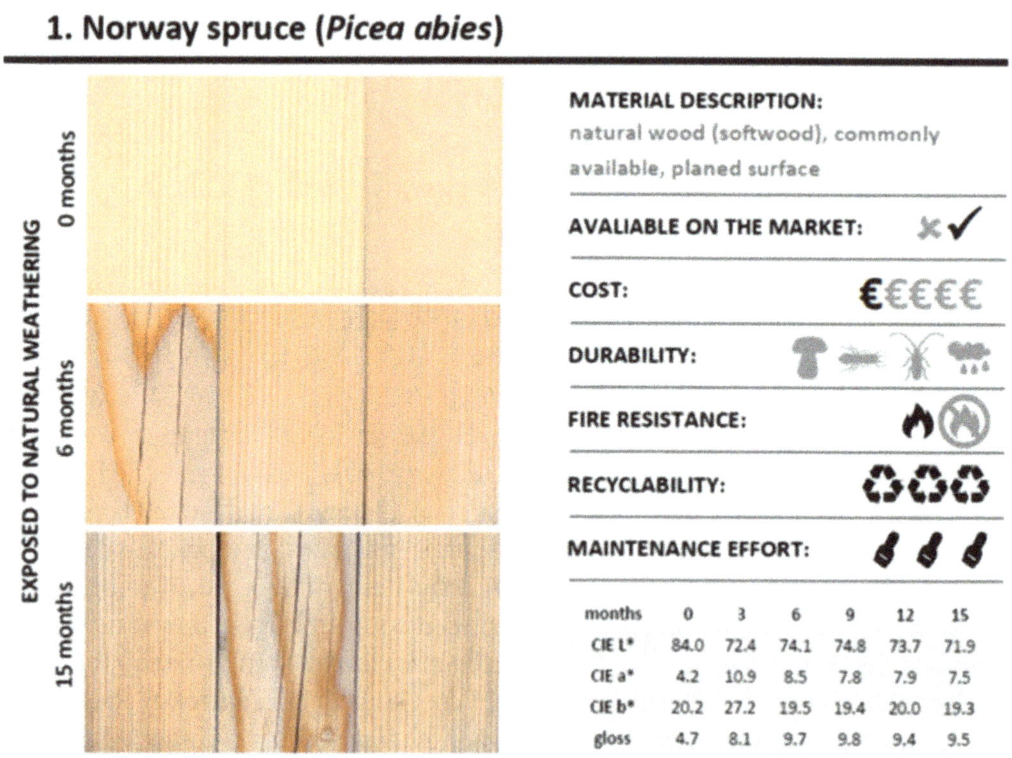

months	0	3	6	9	12	15
CIE L*	84.0	72.4	74.1	74.8	73.7	71.9
CIE a*	4.2	10.9	8.5	7.8	7.9	7.5
CIE b*	20.2	27.2	19.5	19.4	20.0	19.3
gloss	4.7	8.1	9.7	9.8	9.4	9.5

Fig. 6.4 Technical characteristics of Norway spruce wood

2. Sessile oak (*Quercus petraea*)

MATERIAL DESCRIPTION:
natural wood (hardwood), commonly available, sanded surface

AVALIABLE ON THE MARKET: ✓

COST: €€

DURABILITY:

FIRE RESISTANCE:

RECYCLABILITY:

MAINTENANCE EFFORT:

months	0	3	6	9	12	15
CIE L*	69.4	67.5	69.8	73.5	71.4	73.1
CIE a*	6.5	9.2	8.2	6.6	6.7	6.0
CIE b*	20.0	25.6	23.7	20.9	21.1	17.7
gloss	2.7	3.6	4.3	4.9	4.7	5.1

Fig. 6.5 Technical characteristics of Sessile oak wood

3. Teak (*Tectona grandis*)

MATERIAL DESCRIPTION:
natural wood (exotic species from sustainable plantation), planed surface

AVALIABLE ON THE MARKET: ✓

COST: €€

DURABILITY:

FIRE RESISTANCE:

RECYCLABILITY:

MAINTENANCE EFFORT:

months	0	3	6	9	12	15
CIE L*	63.4	72.9	75.2	76.4	77.4	76.2
CIE a*	8.7	7.9	6.5	6.0	5.4	5.8
CIE b*	23.3	20.7	17.2	16.8	15.6	16.8
gloss	5.2	8.2	8.6	7.2	7.4	7.4

Fig. 6.6 Technical characteristics of teak wood

4. Bamboo (*Bambuseae*) siding

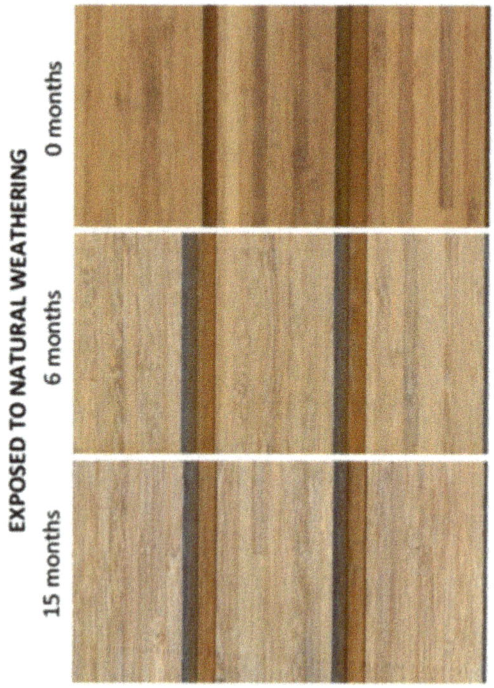

MATERIAL DESCRIPTION:
natural material, glued composite, planed surface

AVALIABLE ON THE MARKET:

COST:

DURABILITY:

FIRE RESISTANCE:

RECYCLABILITY:

MAINTENANCE EFFORT:

months	0	3	6	9	12	15
CIE L*	52.8	62.1	63.5	65.7	64.3	63.4
CIE a*	12.2	10.7	9.0	8.4	7.4	8.1
CIE b*	27.3	29.1	24.5	25.2	22.0	21.1
gloss	2.5	2.8	2.7	2.9	3.0	2.8

Fig. 6.7 Technical characteristics of bamboo

5. Silane impregnated beech (*Fagus* sp.)

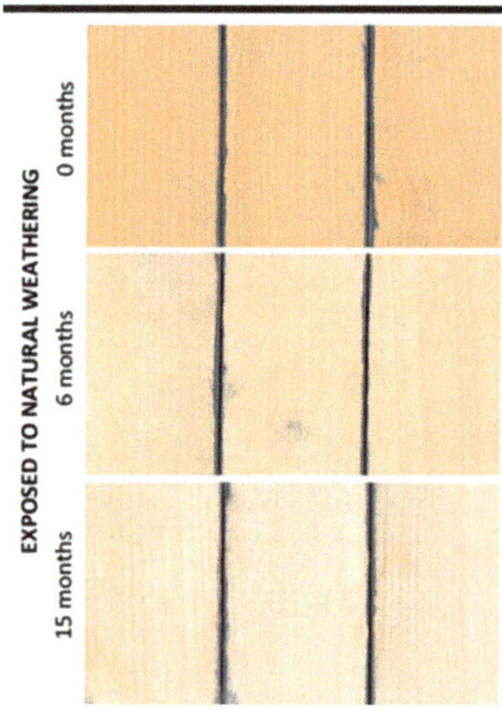

MATERIAL DESCRIPTION:
hardwood impregnated with proprietary silane-based solution

AVALIABLE ON THE MARKET:

COST:

DURABILITY:

FIRE RESISTANCE:

RECYCLABILITY:

MAINTENANCE EFFORT:

months	0	3	6	9	12	15
CIE L*	72.4	73.4	74.8	78.4	80.3	80.4
CIE a*	9.3	9.4	7.5	5.6	4.1	4.5
CIE b*	24.1	25.6	21.9	18.6	16.0	14.3
gloss	3.0	2.8	4.0	3.1	3.0	3.1

Fig. 6.8 Technical characteristics of silane impregnated beech

6. Furfurylated radiata pine (*Pinus radiata*)

MATERIAL DESCRIPTION:
softwood impregnated with furfuryl alcohol

AVALIABLE ON THE MARKET: ✗ ✓

COST: €€€€€

DURABILITY:

FIRE RESISTANCE:

RECYCLABILITY:

MAINTENANCE EFFORT:

months	0	3	6	9	12	15
CIE L*	32.6	45.1	47.9	51.9	52.5	51.5
CIE a*	8.4	6.5	6.8	7.5	7.7	8.2
CIE b*	9.2	17.7	17.6	19.6	19.9	18.6
gloss	6.1	4.5	4.7	4.5	4.2	4.4

Fig. 6.9 Technical characteristics of furfurylated radiata pine

7. Impregnated poplar (*Populus tremula*)

MATERIAL DESCRIPTION:
under utilized hardwood impregnated with proprietary silane-based solution

AVALIABLE ON THE MARKET: ✗ ✓

COST: €€€€€

DURABILITY:

FIRE RESISTANCE:

RECYCLABILITY:

MAINTENANCE EFFORT:

months	0	3	6	9	12	15
CIE L*	82.1	74.2	67.2	73.3	67.5	66.9
CIE a*	8.9	10.6	13.4	9.5	12.4	12.7
CIE b*	35.9	63.6	40.2	31.2	38.5	33.5
gloss	4.3	5.2	4.2	3.9	4.0	3.6

Fig. 6.10 Technical characteristics of impregnated poplar

8. Impregnated Southern yellow pine (*Pinus echinata*)

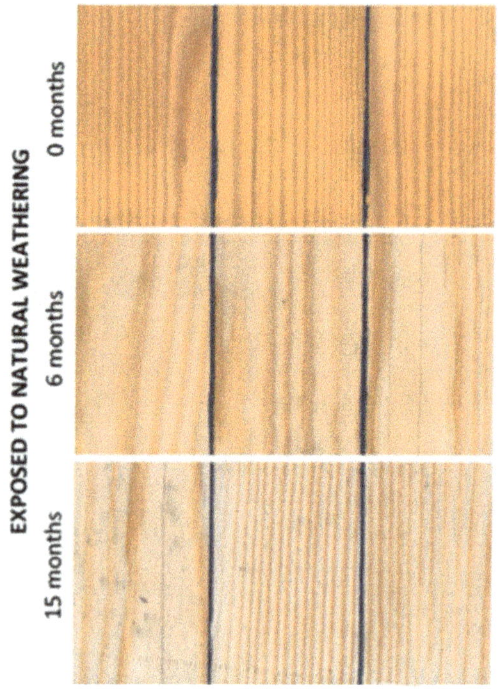

MATERIAL DESCRIPTION:
melamine treated softwood

AVALIABLE ON THE MARKET:

COST: €€€€€

DURABILITY:

FIRE RESISTANCE:

RECYCLABILITY:

MAINTENANCE EFFORT:

months	0	3	6	9	12	15
CIE L*	64.0	66.4	71.4	74.9	73.9	73.5
CIE a*	11.8	11.6	8.4	7.3	6.5	6.7
CIE b*	30.8	30.0	23.1	21.2	19.2	18.7
gloss	4.5	4.3	6.4	9.1	8.0	8.6

Fig. 6.11 Technical characteristics of impregnated yellow pine

9. Thermally modified Scots pine (*Pinus sylvestris*)

MATERIAL DESCRIPTION:
common softwood thermally treated in hot oil

AVALIABLE ON THE MARKET:

COST: €€€€€

DURABILITY:

FIRE RESISTANCE:

RECYCLABILITY:

MAINTENANCE EFFORT:

months	0	3	6	9	12	15
CIE L*	37.5	56.6	67.0	69.4	70.8	71.0
CIE a*	11.1	9.8	6.0	5.3	4.1	3.9
CIE b*	18.0	23.6	20.2	19.4	18.3	16.9
gloss	2.8	5.4	5.0	5.2	5.1	5.4

Fig. 6.12 Technical characteristics of thermally modified pine

10. Thermally modified frake (*Terminalia superba*)

MATERIAL DESCRIPTION:
exotic wood species from sustainable plantation, heat treated in saturated steam

AVALIABLE ON THE MARKET: ✓

COST: €€€€€

DURABILITY:

FIRE RESISTANCE:

RECYCLABILITY:

MAINTENANCE EFFORT:

months	0	3	6	9	12	15
CIE L*	44.7	58.8	65.9	72.3	66.8	67.2
CIE a*	9.8	7.8	5.4	4.2	4.9	4.1
CIE b*	20.0	19.8	14.9	12.8	14.5	11.3
gloss	4.3	3.6	3.7	6.9	4.0	5.

Fig. 6.13 Technical characteristics of thermally modified frake

11. Thermally modified poplar (*Populus tremula*)

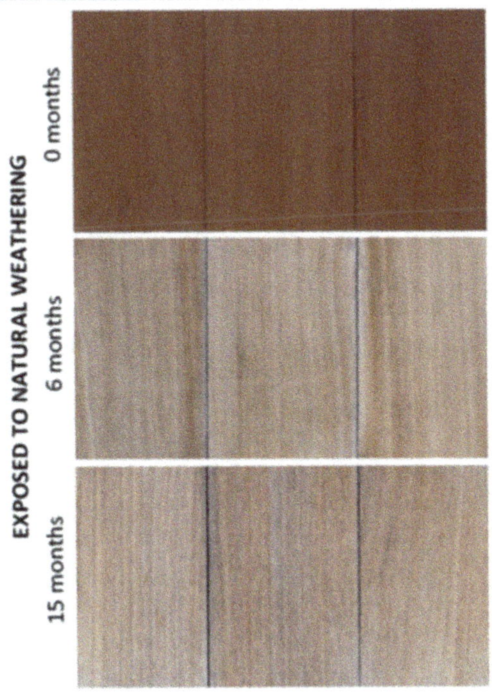

MATERIAL DESCRIPTION:
under utilized hardwood, modified thermally

AVALIABLE ON THE MARKET: ✓

COST: €€€€€

DURABILITY:

FIRE RESISTANCE:

RECYCLABILITY:

MAINTENANCE EFFORT:

months	0	3	6	9	12	15
CIE L*	39.7	59.3	63.3	65.2	64.8	64.3
CIE a*	9.7	6.1	5.0	5.3	5.5	5.2
CIE b*	17.5	17.4	15.3	15.8	15.6	15.0
gloss	2.3	7.1	6.2	7.3	7.2	6.9

Fig. 6.14 Technical characteristics of thermally modified poplar

12. Thermally modified Sessile oak (*Quercus petraea*)

MATERIAL DESCRIPTION:
common hardwood modified thermally in vacuum

AVALIABLE ON THE MARKET:

COST: €€€€€

DURABILITY:

FIRE RESISTANCE:

RECYCLABILITY:

MAINTENANCE EFFORT:

months	0	3	6	9	12	15
CIE L*	55.4	61.5	67.2	69.8	70.6	72.0
CIE a*	6.3	8.1	7.0	5.7	5.5	4.8
CIE b*	17.5	22.1	21.4	19.9	18.9	15.6
gloss	3.7	4.7	5.0	5.0	5.1	4.8

Fig. 6.15 Technical characteristics of thermally modified oak

13. Acetylated alder (*Alnus glutinosa*)

MATERIAL DESCRIPTION:
under utilized hardwood, modified with acetyl anhydrite

AVALIABLE ON THE MARKET:

COST: €€€€€

DURABILITY:

FIRE RESISTANCE:

RECYCLABILITY:

MAINTENANCE EFFORT:

months	0	3	6	9	12	15
CIE L*	57.1	82.8	85.6	85.2	84.5	78.9
CIE a*	7.8	3.5	2.8	2.0	1.4	1.6
CIE b*	19.7	21.9	17.0	14.6	11.6	9.6
gloss	3.4	6.5	6.8	7.9	6.8	7.5

Fig. 6.16 Technical characteristics of acetylated alder

14. Acetylated beech (*Fagus sylvatica*)

MATERIAL DESCRIPTION:
common non-durable hardwood, modified with acetyl anhydrite

AVALIABLE ON THE MARKET:

COST:

DURABILITY:

FIRE RESISTANCE:

RECYCLABILITY:

MAINTENANCE EFFORT:

months	0	3	6	9	12	15
CIE L*	54.6	80.7	81.1	81.3	79.5	68.3
CIE a*	6.9	3.4	3.0	2.4	1.6	1.5
CIE b*	17.7	19.9	17.7	16.0	12.6	9.0
gloss	2.9	4.9	5.0	5.0	4.8	3.9

Fig. 6.17 Technical characteristics of acetylated beech

15. Acetylated radiata pine (*Pinus radiata*)

MATERIAL DESCRIPTION:
softwood, modified with acetyl anhy-drite

AVALIABLE ON THE MARKET:

COST:

DURABILITY:

FIRE RESISTANCE:

RECYCLABILITY:

MAINTENANCE EFFORT:

months	0	3	6	9	12	15
CIE L*	72.0	79.8	81.1	79.8	79.3	76.6
CIE a*	6.3	3.9	2.4	2.2	1.3	1.1
CIE b*	24.1	25.5	16.6	15.5	13.5	9.4
gloss	4.2	5.3	4.7	4.6	6.0	4.9

Fig. 6.18 Technical characteristics of acetylated pine

16. Sessile oak (*Quercus petraea*) finished with wax

MATERIAL DESCRIPTION:
natural wood (hardwood), commonly available, finished with wax

AVALIABLE ON THE MARKET: ✗✓

COST: €€€€€

DURABILITY:

FIRE RESISTANCE:

RECYCLABILITY:

MAINTENANCE EFFORT:

months	0	3	6	9	12	15
CIE L*	67.0	73.4	72.3	76.5	77.4	77.1
CIE a*	7.6	6.9	6.3	5.1	4.5	4.6
CIE b*	32.1	25.4	22.1	22.7	19.5	16.6
gloss	2.2	2.4	2.4	2.8	2.8	2.5

Fig. 6.19 Technical characteristics of oak finished with wax

17. Scots pine (*P. sylvestris*) coated with nanoparticles

MATERIAL DESCRIPTION:
common softwood coated with proprietary nanocoating (TiO_2) solution

AVALIABLE ON THE MARKET: ✗✓

COST: €€€€€

DURABILITY:

FIRE RESISTANCE:

RECYCLABILITY:

MAINTENANCE EFFORT:

months	0	3	6	9	12	15
CIE L*	77.9	69.2	72.4	74.8	73.4	76.0
CIE a*	7.3	11.6	9.3	8.0	7.8	6.5
CIE b*	30.2	30.4	23.2	22.3	22.3	18.1
gloss	3.3	5.2	6.1	4.7	6.7	6.9

Fig. 6.20 Technical characteristics of pine finished with nanocoating

18. Carbonized surface of larch (*Larix* sp.)

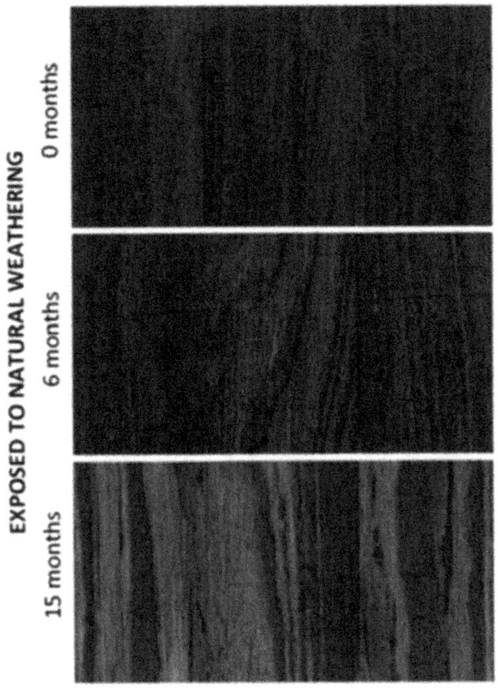

MATERIAL DESCRIPTION:
durable softwood with the surfaces carbonized with fire, finished with oil

AVALIABLE ON THE MARKET: ✗ ✓

COST: €€€€€

DURABILITY:

FIRE RESISTANCE:

RECYCLABILITY:

MAINTENANCE EFFORT:

months	0	3	6	9	12	15
CIE L*	24.8	23.5	26.8	33.2	36.1	32.5
CIE a*	0.1	0.2	0.7	1.5	1.8	1.2
CIE b*	0.3	-0.2	1.0	4.4	6.4	4.2
gloss	1.3	1.3	1.3	2.0	2.4	2.3

Fig. 6.21 Technical characteristics of larch with carbonized surface

19. Pine (*P. sylvestris*) coated with polyurethane paint

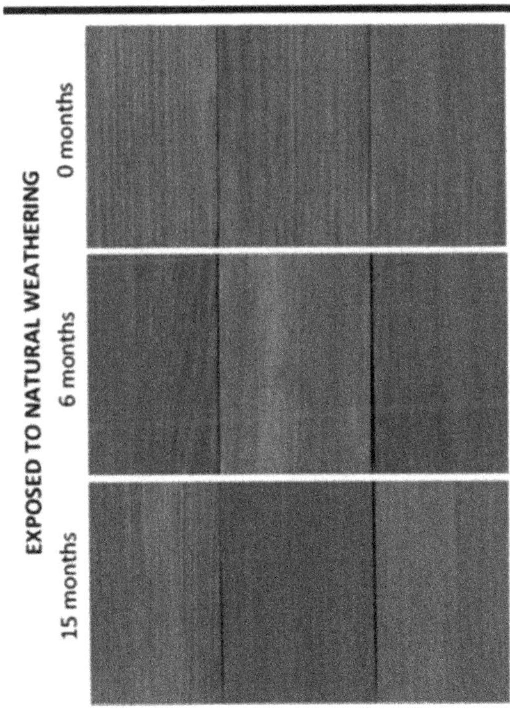

MATERIAL DESCRIPTION:
common softwood coated by hand with DIY polyurethane paint

AVALIABLE ON THE MARKET: ✗ ✓

COST: €€€€€

DURABILITY:

FIRE RESISTANCE:

RECYCLABILITY:

MAINTENANCE EFFORT:

months	0	3	6	9	12	15
CIE L*	51.4	48.7	49.6	49.7	49.6	48.4
CIE a*	5.2	5.5	6.0	5.7	5.5	4.8
CIE b*	13.6	12.3	13.1	12.9	12.4	10.7
gloss	40.3	32.6	27.7	28.9	24.4	21.3

Fig. 6.22 Technical characteristics of pine coated with polyurethane paint

20. TM radiata pine + impregnation with oil

MATERIAL DESCRIPTION:
radiata pine modified thermally + water borne penetrating oil

AVALIABLE ON THE MARKET: ✓

COST: €€€€€

DURABILITY:

FIRE RESISTANCE:

RECYCLABILITY:

MAINTENANCE EFFORT:

months	0	3	6	9	12	15
CIE L*	30.5	38.5	38.3	39.6	39.0	37.7
CIE a*	8.8	10.6	10.0	9.9	9.5	9.0
CIE b*	9.9	16.7	15.1	17.0	16.7	15.5
gloss	1.4	1.6	1.7	1.3	1.3	1.3

Fig. 6.23 Technical characteristics of TM pine impregnated with oil

21. TM radiata pine + impregnation with silicate

MATERIAL DESCRIPTION:
radiata pine modified thermally + a silicate treatment designed to silver off

AVALIABLE ON THE MARKET: ✓

COST: €€€€€

DURABILITY:

FIRE RESISTANCE:

RECYCLABILITY:

MAINTENANCE EFFORT:

months	0	3	6	9	12	15
CIE L*	50.5	68.9	73.0	74.2	76.2	68.7
CIE a*	8.2	4.9	3.9	3.4	2.9	3.6
CIE b*	20.0	14.4	12.1	11.1	10.2	10.7
gloss	1.9	2.7	3.2	2.9	3.1	3.1

Fig. 6.24 Technical characteristics of TM pine impregnated with silicate

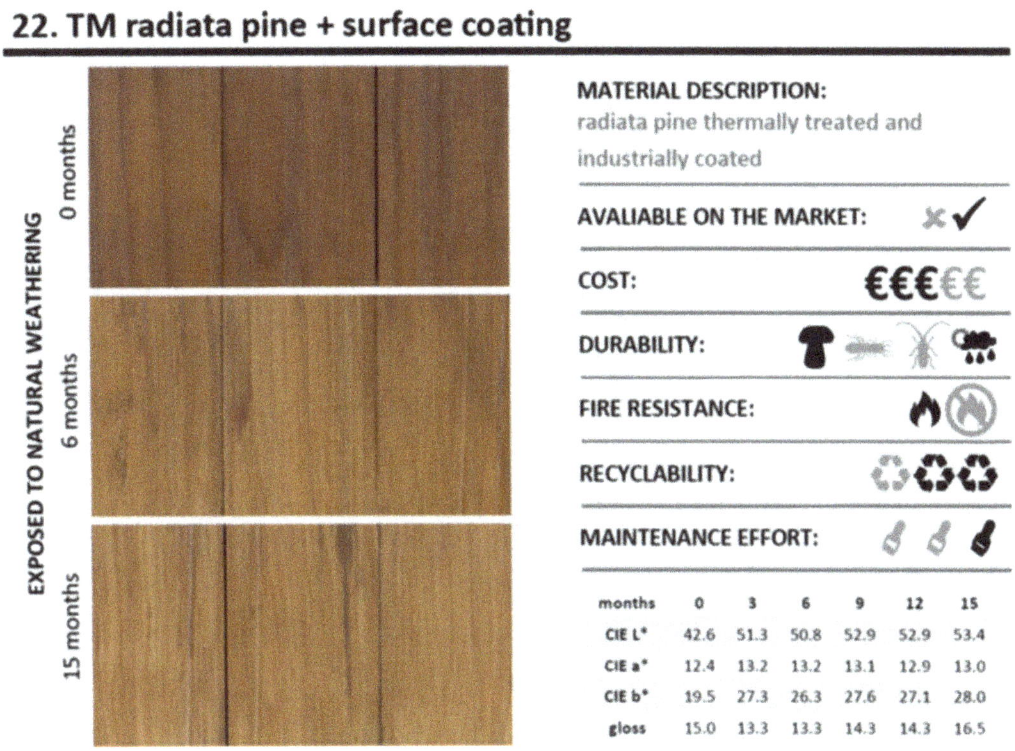

months	0	3	6	9	12	15
CIE L*	42.6	51.3	50.8	52.9	52.9	53.4
CIE a*	12.4	13.2	13.2	13.1	12.9	13.0
CIE b*	19.5	27.3	26.3	27.6	27.1	28.0
gloss	15.0	13.3	13.3	14.3	14.3	16.5

Fig. 6.25 Technical characteristics of TM pine with coated surface

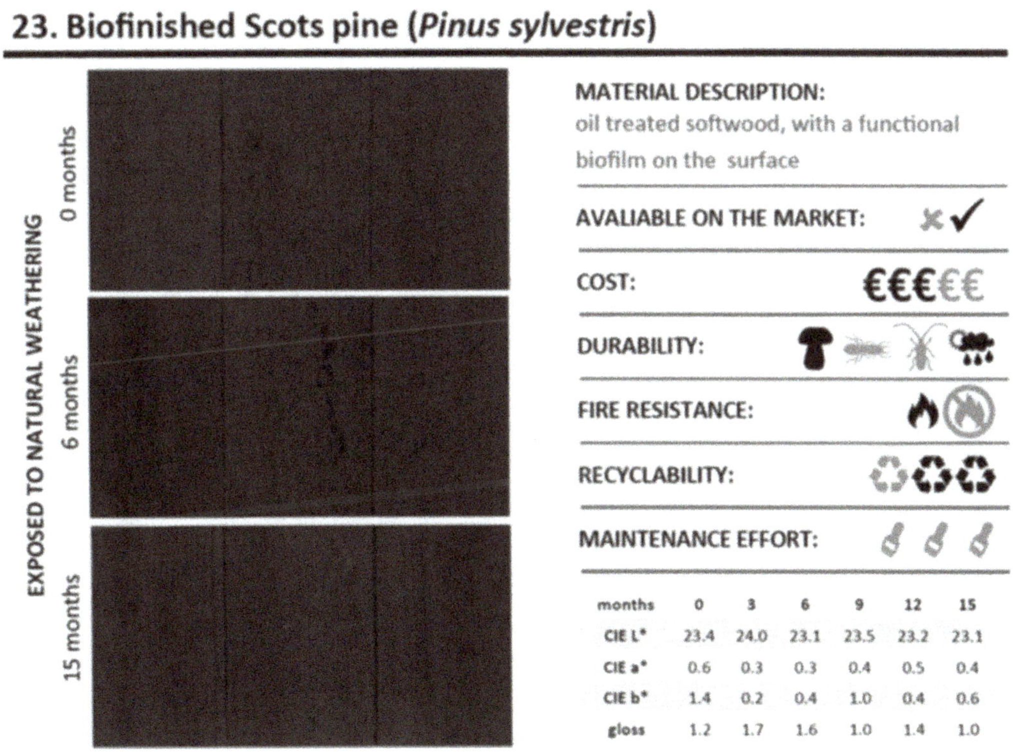

months	0	3	6	9	12	15
CIE L*	23.4	24.0	23.1	23.5	23.2	23.1
CIE a*	0.6	0.3	0.3	0.4	0.5	0.4
CIE b*	1.4	0.2	0.4	1.0	0.4	0.6
gloss	1.2	1.7	1.6	1.0	1.4	1.0

Fig. 6.26 Technical characteristics of biofinished pine

24. Bio-based ceramics

EXPOSED TO NATURAL WEATHERING: 0 months, 6 months, 15 months

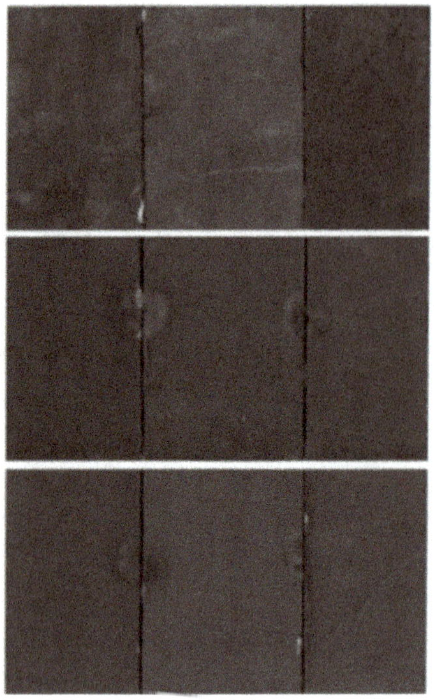

MATERIAL DESCRIPTION:
proprietary composite based on bio-resin

AVALIABLE ON THE MARKET:

COST: €€€€€

DURABILITY:

FIRE RESISTANCE:

RECYCLABILITY:

MAINTENANCE EFFORT:

months	0	3	6	9	12	15
CIE L*	36.8	32.0	35.0	36.9	36.4	36.9
CIE a*	-0.3	-0.1	-0.2	-0.2	-0.1	-0.2
CIE b*	-1.0	-0.5	-0.9	-0.6	-0.5	-0.7
gloss	1.6	1.4	1.0	1.1	1.1	1.2

Fig. 6.27 Technical characteristics of bio-based ceramics

25. Fiberboard

EXPOSED TO NATURAL WEATHERING: 0 months, 6 months, 15 months

MATERIAL DESCRIPTION:
composite board made of straw blended with waste silk fibres + green adhesive

AVALIABLE ON THE MARKET:

COST: €€€€€

DURABILITY:

FIRE RESISTANCE:

RECYCLABILITY:

MAINTENANCE EFFORT:

months	0	3	6	9	12	15
CIE L*	60.8	56.8	58.4	59.4	60.9	56.4
CIE a*	8.3	11.5	9.6	8.6	7.4	8.8
CIE b*	25.8	33.2	30.2	28.2	26.2	26.9
gloss	2.2	1.9	1.7	1.7	1.6	1.3

Fig. 6.28 Technical characteristics of non-wood fibreboard

26. Fibreboard made of acetylated fibres

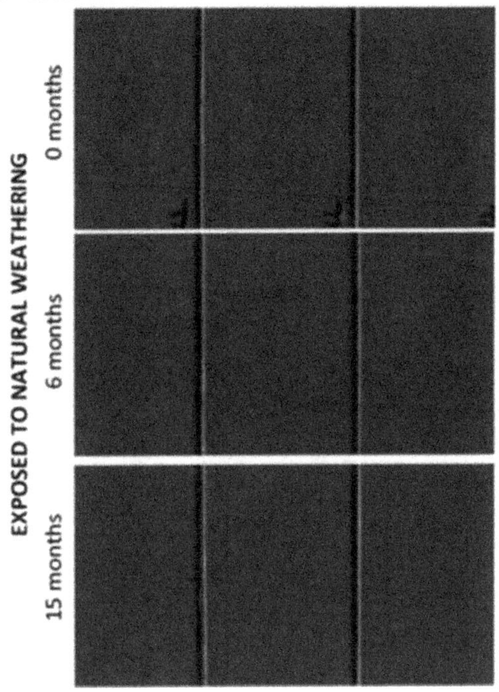

MATERIAL DESCRIPTION:
medium density fibreboard made of acetylated fibres, industrially coated

AVALIABLE ON THE MARKET:

COST: €€€€€

DURABILITY:

FIRE RESISTANCE:

RECYCLABILITY:

MAINTENANCE EFFORT:

months	0	3	6	9	12	15
CIE L*	34.4	34.4	34.5	34.6	34.6	34.4
CIE a*	-2.9	-2.9	-2.9	-2.9	-2.8	-2.8
CIE b*	-2.7	-2.7	-2.7	-2.7	-2.8	-2.7
gloss	80.5	76.3	79.9	80.3	79.3	78.8

Fig. 6.29 Technical characteristics of fibreboard made of acetylated fibres

27. Wood-plastic composite

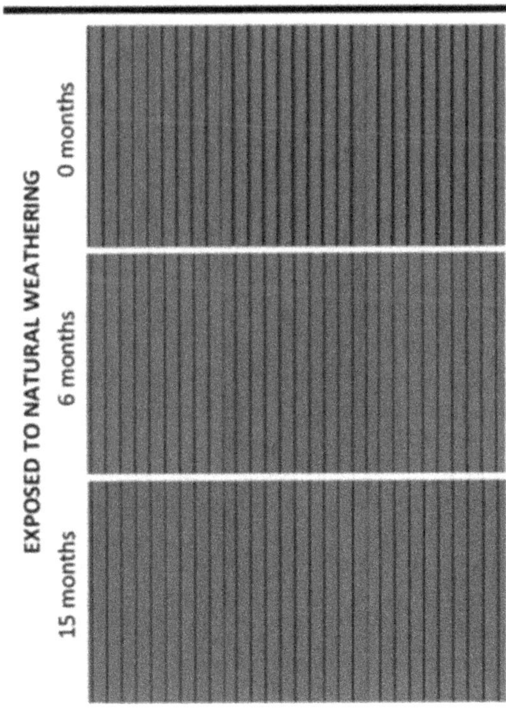

MATERIAL DESCRIPTION:
commercially available wood-plastic composite for general use

AVALIABLE ON THE MARKET:

COST: €€€€€

DURABILITY:

FIRE RESISTANCE:

RECYCLABILITY:

MAINTENANCE EFFORT:

months	0	3	6	9	12	15
CIE L*	50.6	53.4	53.4	54.3	53.0	53.0
CIE a*	-0.4	0.0	0.2	0.2	0.2	0.3
CIE b*	2.2	0.7	0.7	0.7	0.7	0.7
gloss	2.1	2.0	1.8	1.7	1.5	1.6

Fig. 6.30 Technical characteristics of wood–plastic composite

Table 6.2 Explanation of symbols used for sample description

Characteristic	Availability on the market	Cost	Fire resistance	Recyclability potential	Maintenance effort
High (positive)	✓	●●●●●			
Low (negative)	×	○○○○○			

Table 6.3 Explanation of symbols used for assessment of sample durability

Characteristic	Durability			
	Fungi	Termites	Beetles	Weathering
High (positive)			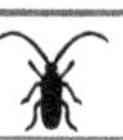	
Low (negative)				

Symbols presented in black mean that material possesses a particular property (e.g., black symbol for termites means that material is termite resistant). Higher number of black symbols in case of cost means that material is more expensive. Similarly, higher number of black symbols in case of recycling means that material has higher recycling potential and can be recycled/reused in several ways.

References

CEN, EN 927-3 (2006) Paints and varnishes—coating materials and coating systems for exterior wood—part 6: exposure of wood coatings to artificial weathering using fluorescent UV lamps and water. European Committee for Standardization, Brussels, Belgium

CEN, EN 927-3 (2014) Paints and varnishes—coating materials and coating systems for exterior wood—part 3: natural weathering test. European Committee for Standardization, Brussels, Belgium

CEN, EN 350-1:1994 (1994a) Durability of wood and wood-based products—natural durability of solid wood—part 1: guide to the principles of testing and classification of the natural durability of wood. European Committee for Standardization, Brussels, Belgium

CEN, EN 350-2:1994 (1994b) Durability of wood and wood-based products—natural durability of solid wood—part 2: guide to natural durability and treatability of selected wood species of importance in Europe. European Committee for Standardization, Brussels, Belgium

CEN, EN 252:2014 (2014) Durability of wood and wood-based products. field test method for determining the relative protective effectiveness of a wood preservative in ground contact. European Committee for Standardization, Brussels, Belgium

CEN, EN 350:2016 (2016) Durability of wood and wood-based products—testing and classification of the durability to biological agents of wood and wood-based materials. European Committee for Standardization, Brussels, Belgium

Jones D, Brischke C (eds) (2017) Performance of bio-based building materials. Woodhead Publishing

Kutnik M, Suttie E, Brischke C (2014) European standards on durability and performance of wood and wood-based products—trends and challenges. Wood Mater Sci Eng 9(3):122–133
Kutnik M, Suttie E, Brischke C (2017) Durability, efficacy and performance of bio-based construction materials: standardisation background and systems of evaluation and authorisation for the European market. In: Jones D, Brischke C (eds) Performance of bio-based building materials. Woodhead Publishing, pp 593–610
Reinprecht L (2016) Wood deterioration, protection and maintenance. Wiley, Oxford, UK

Chapter 7
Future Perspectives

Logic will get you from A to B. Imagination will take you anywhere.

Albert Einstein.

Abstract This chapter concludes the overall content of this book. The key challenges regarding possible innovations in the building sector are presented from the perspective of the development in materials science, design concepts, as well as tools and services for improved façade management.

The main purpose of a building envelope is to protect occupants from the weather and provide a basic shelter. Nowadays, building façades perform many more functions than in the past, offering security, privacy, comfort, as well as benefits such as aesthetic pleasure and improved well-being. A building façade reflects civic and cultural identity; therefore, we want our homes to be unique and individualized. In fact, façade design can make buildings become iconic and be remembered for centuries. On the other hand, we want to perceive façades as honest and sincere.

The key challenge in designing building façades is meeting the occupant needs while reducing building energy consumption, according to requirements of global policies (OECD/IEA 2013; EPBD 2018). The modern architecture focuses on every stage of the building process. It starts with the design and planning phase, includes construction and building performance, as well as the disposal of building materials. The way building façades are designed is changing. Besides waterproofing, insulation, and aesthetic functions, façade planning includes energy efficiency, which adds another layer of complexity to the design process. Forthcoming developments regarding building façades are related to the innovation of materials, alternative design, and facilities enabling essential functioning and operation.

Building envelopes should guarantee functional and aesthetic performance. Technological development is constantly bringing to the market materials enriched by new technologies that can reduce the environmental impact, while improving health and safety. New materials (both cladding and insulation) and construction methods (prefabrication and modular system) enable a fast and accurate construction process. Bio-based products have the advantage of a significantly lower

A. Sandak et al., *Bio-based Building Skin*, Environmental Footprints and Eco-design of Products and Processes, https://doi.org/10.1007/978-981-13-3747-5_7

embodied carbon manufacturing profile than steel or concrete, with the added benefit of carbon sequestration. Mass timber in tall building applications is under investigation as an alternative to steel and reinforced concrete structural systems, with feasible options already at a height of 40 stories and above (Larasatie et al. 2018). Using bio-based building materials in the building sector moves the traditional building concept towards the green architecture. In fact, green buildings have become one of the most important and progressive trends in the building industry during the last 20 years (Jadhav 2016). In this perspective, environmentally friendly materials originating from plants or animals (e.g., wood, bamboo, cork, hemp, flax, straw, sheep wool, seaweed, mushrooms hyphae) are in great interest of architects.

Sustainable construction principles require the use of local materials and products. An example might be the Living Building Challenge certification programme and sustainable design framework that require a certain percentage of building materials to be originated from within a certain distance to the construction site (Brown 2016). Due to their versatile character, bio-based materials enable creating buildings adapted to any local context. Different natural materials usually look good together even if having different form or patterns (Kotradyová and Teischinger 2014). Thus, biomaterials give an impression of being suitable for any context. Examples of a rustic and modern architecture being in synergy with the environment are presented in Fig. 7.1.

As static structures, traditional buildings remain the same for decades, interacting with their occupants and the environment in limited ways. Sustainable design aims to create buildings that respond to the environment and react to climate changes. In addition, it focuses on living conditions, resulting in regenerative, restorative, and adaptive buildings. Restorative architecture is the next stage of green architecture. Here, a building gives more to the environment during its lifetime than it takes away during the construction. Instead of focusing narrowly on the building design, such buildings tend to integrate with the natural environment while being environmentally friendly. When optimizing orientation, shape, and layout of buildings, we should consider both, needs of society and relationship with the environment. Humans are inherently drawn to nature; however, urban

Fig. 7.1 Wooden buildings designed in rustic and modern style well fitted in mountain landscape Photograph courtesy of Paolo Grossi (left) and Accsys (right)

Fig. 7.2 Different approaches for connection with nature. Duna Building (Bergen School of Architecture, Slovak University of Technology in Bratislava), photograph courtesy of Veronika Kotradyová (left) and Romanian Pavilon at the World Expo 2000, arch. Doru O. Comsa, 2000 (right)

environments are disrupting this relationship. Biophilic design encourages this affinity by creating natural environments for living, working, and learning. By consciously including natural materials in interior or architectural design, we are unconsciously reconnecting to nature (Fig. 7.2).

According to the World Health Organisation, stress-related illnesses (e.g., mental health disorders and cardiovascular disease) will be the largest causes by 2020 (Vigo et al. 2016). Using wood as an exposed material in buildings where humans can interact with it is known to create positive psychophysiological effects for building users. Incorporating nature into the built environment, either directly (e.g., potted plants) or indirectly (e.g., tree-like columns), can reduce physiological and psychological indicators of stress, while increasing productivity, creativity, and self-reported levels of well-being (Mcsweeney et al. 2015). Research in this area provides an evidence base of positive health impacts of wood use in the built environment (Burnard and Kutnar 2015). One emerging area in this research field is Restorative Environmental and Ergonomic Design (REED), a building design paradigm that can provide guidance for wood use in buildings to improve human health (Burnard et al. 2016). Integrating nature into the built environment, by enabling views of nature, by using natural materials (especially local materials), and by reflecting local ecology in building design and use, is thought to improve building users' perception of the natural environment and therefore motivate them to care for it (Derr and Kellert 2013). REED is integrating frameworks for improving occupant and user health, increasing safety, and improving building management. This represents a shift in building design (and neighbourhood/community design) from minimizing environmental harm towards creating positive impacts for the natural environment, building users, and society in general.

Bio-based building materials fit very well with the general concept of minimizing the amount of waste based on the "reduce–reuse–recycle" paradigm. Reused engineered solid wood products are a highly valuable source of construction materials following cascade use principles. Moreover, such architecture design directs public attention towards the benefits of a low-waste, circular economy. Material reuse and recover are linked not only to the recyclability of individual products but also to the construction techniques that enable fast disassembly (Casini 2016). Sustainability in this perspective is achieved by using smarter materials, optimizing design and layout, and reducing long-term maintenance costs and recovery of materials at the end of their service life. Therefore, durability, sustainability, and low-maintenance will continue to be factors that influence the selection of exterior materials.

Innovation in architectural design requires applying new design strategies and new project delivery methods. Progress in information technology, sensors, and computation transforms design processes and procedures, delivery methods, as well as fabrication and construction methods (Aksamija 2016). Computational performance-based design, building information modelling (BIM), simulations, virtual construction, and digital fabrication are transforming contemporary architecture practice. Building information modelling (BIM) is a design procedure that includes the administration/management of the digital representation of physical objects composing the planned building (structure, sub-structure, installations, and systems). The general idea resembles the concept of the "Internet of Things", where every physical object has its virtual representation, with the difference that BIM is used to design and plan buildings. BIM files enable storing and exchanging the information, and—what is the most important feature—the networking on the project. BIM systems are utilized by individual designers, organizations, and government offices. In the context of material ageing, BIM is a very powerful tool that makes it possible to examine potential changes in material properties and appearance and plan future maintenance. Ongoing research aims to expand BIM specifications to include object performance data, environmental impact data, and health impact data, among other expansions. Consequently, new functionalities, such as 4D (construction sequencing), 5D (costs), 6D (sustainability), and 7D (facility management), provide new complementary extensions of the BIM utilities.

Progress in material science and new concepts in the building design phase create a vast amount of multidisciplinary technological knowledge regarding building façades. The development of new technologies and concepts should be supported with multi-scale characterization methods at a component level. This should be followed by whole-life evaluation methods at the building level and building/user integration. The most promising concepts should be identified and disseminated between the different stakeholders in the façade industry and the academia. Finally, lifelong training programmes for future generation of architects and designers should be implemented in order to bring new design concepts into existence.

References

Aksamija A (2016) Integrating innovation in architecture. In: Design, methods and technology for progressive practice and research. Wiley, Chichester

Brown M (2016) Futurestorative. In: Working towards a new sustainability. Riba Publishning, Newcastle

Burnard MD, Kutnar A (2015) Wood and human stress in the built indoor environment: a review. Wood Sci Technol 49(5):969–986

Burnard MD, Kutnar A, Schwarzkopf M (2016) Interdisciplinary approaches to developing wood modification processes for sustainable building and beyond—InnoRenew CoE. In: Proceedings of the world conference on timber engineering, Vienna, Austria, pp 348–354

Casini M (2016) Smart buildings: advanced materials and nanotechnology to improve energy efficiency and environmental performance. Woodhead Publishing

Derr V, Kellert SR (2013) Making children's environments "R.E.D.": restorative environmental design and its relationship to sustainable design. In: Pavlides E, Wells J (eds) Proceedings of the 44th annual conference of the environmental design research association. Providence, Rhode Island

EPBD (2018) Energy performance of buildings directive. European Commission 2018/844/EU

Jadhav NY (2016) Green and smart buildings. In: Advanced Technology Options. Springer Nature Singapore

Kotradyová V, Teischinger A (2014) Exploring the contact comfort within complex well-being in microenvironment. Adv Mater Res 899:356–362

Larasatie P, Guerrero JE, Conroy K et al (2018) What does the public believe about tall wood buildings? An exploratory study in the US Pacific Northwest. J For 116(5):429–436

Mcsweeney J, Rainham D, Johnson SA et al (2015) Indoor nature exposure (INE): a health-promotion framework. Health Promot Int 30(1):126–139

OECD/IEA (2013) Technology roadmap. In: Energy efficient building envelopes. International Energy Agency, France

Vigo D, Thornicroft G, Atun R (2016) Estimating the true global burden of mental illness. Lancet Psychiatry 3:171–178

www.ingramcontent.com/pod-product-compliance
Lightning Source LLC
LaVergne TN
LVHW082233200926
848484LV00008B/2350
* 9 7 8 1 0 1 3 2 7 0 6 3 5 *